T0291776

Recent Trends in Computer-aided Diagnostic Systems for Skin Diseases

Recent Trends in Computer-aided Diagnostic Systems for Skin Diseases

Theory, Implementation, and Analysis

SAPTARSHI CHATTERJEE
Department of Electrical Engineering, Jadavpur University,
Kolkata, India

DEBANGSHU DEY
Department of Electrical Engineering, Jadavpur University,
Kolkata, India

SUGATA MUNSHI
Department of Electrical Engineering, Jadavpur University,
Kolkata, India

Academic Press is an imprint of Elsevier
125 London Wall, London EC2Y 5AS, United Kingdom
525 B Street, Suite 1650, San Diego, CA 92101, United States
50 Hampshire Street, 5th Floor, Cambridge, MA 02139, United States
The Boulevard, Langford Lane, Kidlington, Oxford OX5 1GB, United Kingdom

British Library Cataloguing-in-Publication Data
A catalogue record for this book is available from the British Library

Library of Congress Cataloging-in-Publication Data
A catalog record for this book is available from the Library of Congress

ISBN: 978-0-323-91211-2

For Information on all Academic Press publications
visit our website at https://www.elsevier.com/books-and-journals

Publisher: Mara Conner
Acquisitions Editor: Sonnini Yura
Editorial Project Manager: Zsereena Rose Mampusti
Production Project Manager: Poulouse Joseph
Cover Designer: Matthew Limbert

Typeset by MPS Limited, Chennai, India

Dedicated to -

Mr. Shyamal Chatterjee, Mrs. Papri Chatterjee and also to

Mrs. Priyanka Saha Chatterjee

&

Late N. C. Dey, Mrs. Sabita Dey and also to

Mrs. Tanaya Dey, Ms. Aratrika Dey

&

Late Suprasanna Munshi, Late Anjali Munshi and also to

Mrs. Kakali Munshi, Ms. Anwesha Munshi

for being the strength, inspiration and support behind each word...

Contents

Preface

The early and accurate diagnosis of diseases is of uttermost importance for the timely and proper medical intervention, and skin diseases are no exception. For a long time, dermatologists and other medical personnel relied on visual examination of the skin or relevant images and the domain expertise amassed over decades to diagnose the type of skin disease and its prognosis. In many cases, it is very difficult for the medical experts to provide accurate and uniform evaluation of some skin diseases from visual impressions. It was therefore quite natural that with the advent of processor-based systems, and with the rapid strides made in their improvement as well as dwindling prices, computer-aided medical diagnosis systems would be developed and the deployment of these systems would become a part and parcel of medical diagnostic protocols.

The unaided clinical examination of pigmented skin lesions, which has limited and variable identification accuracy, leads to important challenges in early detection of disease and consequent minimization of unnecessary biopsy. Visual similarities in the manifestation of different skin lesions in terms of morphological, textural, and color complexity makes early diagnosis a difficult task for both general physicians and dermatologists. The basic framework of a typical computer-aided diagnostic (CAD) system for skin disease monitoring consists of an image preprocessing module, segmentation, feature extraction, and a feature selection module, followed by a classification module.

The present book is an attempt to elaborate on the traits of state-of-the-art techniques in CAD systems for dermatological diseases. The book is meant for experts as well as uninitiated research personnel preparing for a successful journey in the domain under consideration.

The organization of the book is as follows:

Chapter 1, Introduction, presents a brief introduction on the importance of modern CAD techniques for skin disease identification as well as a glimpse of different noninvasive and noncontact monitoring systems.

Chapter 2, Preprocessing and Segmentation of Skin Lesion Images, deals with preprocessing techniques and their importance for dermoscopic images, followed by segmentation of skin lesions.

Chapter 3, Extraction of Effective Hand Crafted Features From Dermoscopic Images, discusses different digital signal processing tools for the extraction of effective hand-crafted features from dermoscopic images.

Chapter 4, Feature Selection and Classification, addresses feature selection techniques to obtain most significant and differentiating features for various skin disease classes and classification models for binary and multiclass classification problems.

In Chapter 5, Development of Expert System for Skin Disease Identification, a step-by-step method for developing an integrated expert system implementing knowledge to replicate gold standard rule of dermoscopy is elaborated.

Finally, Chapter 6, Conclusions and Future Scope of Work, draws conclusions with ideas about the future scope of research in dermatological diagnostic systems.

The authors are grateful to the Electrical Engineering Department, Jadavpur University, for providing the laboratory facilities required for carrying out the research required for this book. The authors would like to extend their sincere gratitude to Dr. Surajit Gorai, consultant dermatologist, Apollo Gleneagles Hospital, Kolkata, India, for his valuable suggestions. The authors will always remain indebted to him for taking time out of his busy schedule and sharing his medical expertise. The authors express their thanks to the faculty members and research scholars in the Electrical Engineering Department of Jadavpur University for their constructive criticism and useful suggestions. The authors sincerely thank their families for immense support and cooperation during the writing of this book.

The authors would like to acknowledge Visvesvaraya PhD Scheme, MeiTY, Government of India, for providing financial support through Jadavpur University to pursue research work in a smooth and hassle-free manner.

Finally, responsibilities for any mistakes and for the ideas expressed in this book are those of the authors alone.

Saptarshi Chatterjee
Debangshu Dey
Sugata Munshi

CHAPTER 1

Introduction

1.1 Background

Human physiology mainly deals with different biological subsections including organs, cells, biological compounds, and also focuses on their functions or interactions to make life possible. From ancient theories tomolecular laboratory techniques, physiological research has modulated the understanding of the components of human body, and their way of communication to keep us alive. Physiology examines how organs and systems within the human body function and combine their efforts to make conditions favorable for survival [1].

Some of the major subsystems of the human physiological system are as follows:

- *Circulatory system*: includes the heart, blood, blood vessels, the blood circulation during both healthy, and unwell situations.
- *Digestive system*: includes the spleen, liver, and pancreas, the conversion of food into energy and the final exit of the excreta from the body.
- *Respiratory system*: consists of noses, nasopharynx, trachea, and lungs. This system brings in oxygen and expels carbon dioxide from the body.
- *Integumentary system*: consists of the skin, hair, nails, sweat glands, and sebaceous glands.
- *Nervous system*: consists of central nervous system (human brain and spinal cord) and the peripheral nervous system.

The conditions and activities of these physiological systems manifestas different biochemical, biomechanical, and bioelectrical (also called biomedical) signals. Diverse forms of biomedical signals contain a wide variety of information of these physiological processes. Abnormal conditionor ill-health of the associated components of any physiological system is recognized from the corresponding biomedical signals. The processing and in-detailed study on such biomedical signals is directed toward the assessment of the state of the system [2]. Analysis and interpretation of a biomedical signal by medical personnel bear the weight of the experience and expertise of the analyst; however, such analysis is subjective in nature. The visual inspection of these physiological abnormalities may sometimes lead to the improper diagnosis at an early stage of the disease. Today, the development of digital signal processing tools and computer vision algorithms are playing an important role in early-stage detection of the diseases from the biomedical signals.

Recent Trends in Computer-aided Diagnostic Systems for Skin Diseases
DOI: https://doi.org/10.1016/B978-0-323-91211-2.00004-4

1.1.1 Condition monitoring of biomedical systems

The information regarding the condition of various biomedical systems can be captured through different biomedical measurement systems. The term measurement generalizes the acquisition of various clinical information in the form of biomedical signals or images related to the anatomical sites. To potentially determine the patients' health, different non-invasive and non-contact measuring techniques are developed [3,4]. The acquisition of biomedical signals like electroencephalogram (EEG), respiration, electromyogram, electroculogram, oxygen saturation level (SpO2), and electrocardiogram (ECG), and so on, provide essential information to the clinicians to assess the physical state of the patient. Biomedical imaging focuses on the acquisition of images for both diagnostic and therapeutic purposes. Advanced measurement and sensing technologies with modern computing techniques are used to develop image acquisition modules to garner sufficient information about the physiology or physiological processes [5]. Biomedical image acquisition tools are developed by employing X-rays in computed tomography (CT) scans, ultrasound in ultrasonography (USG), electromagnetism in magnetic resonance imaging (MRI) or light in endoscopy, optical coherence tomography (OCT), dermatoscopy (or dermoscopy), and so on to judge the condition and further monitoring of associated organs or tissues of different biological systems. Biomedical signal processing tools involve the analysis of these measurements using various mathematical formulae and algorithms to provide more insights to aid in clinical assessment and condition monitoring of the systems. So, a typical biomedical condition monitoring system comprises the following functional components [6]:

- *Sensor*: The *sensors* sense the physiological activities of various systems of human body and convert them to electrical signals. Biosensors consist of primary sensing elements and variable conversion elements. Sensors respond to various physical measurands like biopotential, pressure, flow, dimension (imaging), and so on and convert them to suitable electrical form. A signal conditioning circuitis associated with the biomedical sensors to amplify and filter the signal and subsequently converts it to digital form for further processing and storage. On the basis of the integrated transducer technology, biosensors can be classified as electrochemical, potentiometric, impedence spectroscopy based, piezoelectric, and so on [6,7].
- *Processing module*: The *processing module* is responsible for developing efficient signal or image processing algorithm for the extraction of meaningful and clinically correlated information from the acquired biomedical signal. Digital signal processing tools are developed to extract different statistical and clinical data regarding the physiological phenomena [7]. The signal processing tools help in identification and further analyses of the revealing information obtained from the acquired signals, such as variations of P, Q, R waves of ECG signal, presence of nodule in lung, widened or ruptured blood vessels, and so on. The analysis reveals the reasons

behind the abnormal behavior of the associated physiological system, necessitating the continuous monitoring of the disease.

- *Display module*: The *display module* interfaces the biomedical condition monitoring system with the expert. The display module helps the doctors in visual inspection of different biological parameters of associated physiological system to make further decision. Condition monitoring system portrays the biological parameters or signals or different anatomical sites by numerical, graphical, or visual means. The ECG, SpO2 signals, and so on portray the heart potential and the oxygen saturation levels. Quantitative values of pulse rate and blood pressure help to monitor the physical health of the patient. However, to detect the rupturing of blood vessels or abnormal growth of tissue or pigmentation in different locations of human body, radiologists use different imaging modalities to cross-verify the physiological abnormalities [7].

1.1.2 Non-invasive and non-contact monitoring techniques for different biomedical systems

Non-invasive procedure is a conservative diagnostic or therapeutic approach, where no break in the skin is created and no contact with the mucosa is required. During the non-invasive monitoring, the patient is not given any drug by any means, orally or by injecting. In medical science, non-invasive methods encompass simple observations to aiding specialized forms of surgery. Non-invasive imaging with signal acquisition techniques are widely used for the identification and monitoring of different biomedical abnormalities. From the wide spectrum of non-invasive diagnostic images, doctors and radiologists use dermatoscopy, CT, MRI, OCT, USG, and so on for the in-depth visualization of anatomical and physiological processes of human body.

The physiological abnormalities at different anatomical sites of the human body are not always accessible to the radiologists. Locations of various physiological disorders need to be assessed with more in-depth visualization. Depending on such requirements for early and accurate diagnosis, radiologists use different non-invasive image acquisition techniques with advanced sensing technology. Recent technological development has lead to the conversion and storage of such acquired signal in form of high-resolution digital images. Different noninvasive image acquisition techniques are discussed here with some applications.

1.1.2.1 Computed Tomography (CT) scan

The limitations of conventional two-dimensional X-ray imaging in the detection of small variations is circumvented by the development of CT. In this imaging technique, the structures lying in narrow anatomical slices are digitally reconstructed by projecting collimated beam of X-rays through different orientations until an angle of 180° is swept. CT images help to detect less-dense-tissue information from dense bony structures. CT images with large number slices help the radiologists to identify and monitor the tumors, metastatis, and defects in blood vessels [6].

1.1.2.2 Magnetic Resonance Imaging (MRI)

The non-invasive nuclear magnetic resonance signals are used to produce the image in MRI technique. Similar to CT scan, in MRI, multiple line signals are processed with incremental perturbation of magnetic field to produce small magnetic gradient. Unlike CT scan, the entire setup is not rotated; rather the direction of magnetic gradient is rotated slightly [7]. The MRI images help to obtain the density profile or contrast detail of soft tissues of human body with higher resolution than CT images. MRI images are used for in-depth visualization of anatomical structures like gray and white matter of brain, small cancerous lesions in liver, and so on.

1.1.2.3 Ultrasonography

USG is an ultrasound based diagnostic imaging technique used to visualize subcutaneous structures of human organs like abdomen, blood vessels, female reproductive organs, heart, and so on. The rules of propagation and reflection of ultrasound are used to develop ultrasound scanner, utilizing the piezoelectric properties of ultrasound transducer. Unlike the ionizing radiation of X-rays, ultrasound energy is not harmful for the patients. Here, the fine structures and locations of the tissues are obtained from the time delay between the transmitted pulse and its echo. Most of the medical USG arrangements use ultrasound frequencies from 1.0 to 15 MHz [7].

1.1.2.4 Optical coherence tomography

OCT is a non-contact, non-invasive imaging technique employing low coherence interferometry based on near-infrared light source, to acquire images of biological tissues. OCT imaging technique acquires images with submicrometer resolution owing to the use of broad-bandwidth light source, in comparison to high frequency ultrasound imaging which is limited to depths of a few millimeters. Similarly, transverse resolution for ultrasounds is lower than the OCT. Recent developments in OCT technology introduce high resolution and ultra-high resolution imaging employing laser light source with axial resolution of 1 μm. OCT has the highest clinical impact in the study of opthalmology. The OCT image portrays a high-resolution cross-sectional view of human retina and is used to detect and monitor glaucoma, chorioretinopathy, macular hole, optic disk pits, and so on. In medical application, OCT imaging is now effectively used in the identification of early neoplastic changes [7,8].

1.1.3 Automatic disease monitoring systems

The indispensable role of digital medical imaging in modern healthcare has accelerated the development of advanced signal processing tools for the analysis and further decision making. The biomedical signals, acquired using non-invasive techniques are used by the doctors and the radiologists for the identification of the physiological abnormalities from visual inspection only. The early and accurate diagnosis leads to the proper treatment and prevention of the disease. However, very little change in visual appearance at the early stage of the disease, makes it difficult for clinical practitioner to differentiate any such change. Sometimes it becomes very difficult to analyze a large number of patients' data in restricted span of time. Here, a computer-aided diagnostic (CAD) system plays an important role in analyzing the patients' data and further decision making in a restricted span of time. Diagnosis of the disease using computer assisted decision making system will work well to generate a second opinion, apart from the human expert's opinion. A typical computer-aided disease monitoring system comprises the following major subsystems, as depicted in the block schematic representation in Fig. 1.1.

1.1.3.1 Pre-processing

Biomedical signals are usually low voltage signals that may be contaminated by noise and artifacts. Prior to the signal processing stage, the acquired biomedical signal is passed through a pre-processing stage to eliminate the effect of unwanted interference of noise. The contaminating noise in biomedical signals originate from four different types of sources, namely physiological variability, intrinsically generated noise (e.g., in transducers) or environmental noise or interference. Physiological noise occurs due to constant movement of the patient, unusual breathing, body temperature, and so on. The external interference like 50 Hz power supply interference, insufficient or uneven illumination of light, and so on introduce noise in acquired biosignals or images. The sensor nonlinearity or sensor saturation also introduce noise in the raw biomedical signals. Literature suggests different noise removal techniques for wide variety of biomedical signals [9–12]. Ramakrishnana et al. have employed high-pass filter to eliminate baseline wander, and high frequency noise of ECG signal in [13]. Nayak et al. have introduced a robust digital fractional order differentiator-based preprocessing technique for better QRS detection accuracy from ECG signal [14]. Biomedical images often contain unwanted artifacts depending on the anatomical sites of the affected area. The image pre-processing techniques are widely used in biomedical image processing applications in order to improve the quality of the images for further processing.

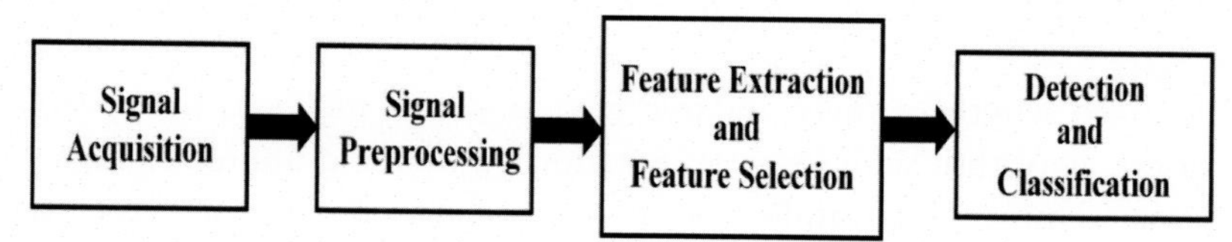

Figure 1.1 Block diagram of a typical automatic disease monitoring system.

In [15], Rai and Chatterjee have proposed discrete wavelet transform-based independent component analysis technique to eliminate noises in MRI images. For the segmentation of vessels from retinal fundus images, Pal et al. have used 2D wavelet transform and Gaussian filtration followed by morphological operations to identify the vessels more prominently and to exhibit enhanced vessel background contrast [16]. In literature, more such pre-processing techniques are reported that use digital filters and advanced image processing algorithms to eliminate the unwanted noise and artifacts without altering the useful information [17–19].

1.1.3.2 Feature extraction

The pre-processed signal is manipulated in the subsequent stages for the extraction of meaningful information. Finding well representative data is very much domain specific. For these modules, digital signal processing tools are widely explored to quantify the representative features or clinical findings, inaccessible by visual inspection by the human expert. Feature extraction methodologies encompass different traditional transformation and non-transformation techniques for the quantification of statistical, structural, textural or other high level features of the biomedical signals. Koley and Dey have extracted a set of time domain and frequency domain features along with some nonlinear features like entropy, fractal dimension and so on from the EEG signals to detect apnea and hypopnea event for portable sleep apnea monitoring device [20]. Mar et al. have explained ECG beats classification technique using temporal, morphological and statistical features [21]. In order to identify various physiological phenomenon for 1D biomedical signals, different state-of-the-art feature extraction tools have been discussed in literature [22–24]. For the identification of different diseases from biomedical images, extraction of morphological, textural and color features have been reported [18,25,26]. Correlation based block matching technique has been performed for the registration of CT to ultrasound volumes for image guided liver interventions by Banerjee et al. [27]. Ram et al. have proposed clutter-rejection-based approach using distance and correlated features to extract microaneurysms from digital fundus images for the early detection of diabetic retinopathy [28]. These extracted quantitative features are considered to be the representatives of the physiological conditions. From this large feature set, identification of the most compact and demarcating features is important for improved classification of the abnormality. Feature selection techniques are introduced to obtain a reduced set of features with the following advantages.

1. The reduced length of representative feature vector increases the speed of execution of the algorithm with low computation burden and limited storage.
2. Selection of most demarcating features differentiate the disease classes with higher degree of accuracy.

1.1.3.3 Classification

In the classification stage, the input signal is categorized in representative classes on the basis of extracted features. In the final stage of the automatic disease monitoring system, the physiological disorder is identified and correlated with the clinical symptoms. According to the various physiological phenomena, the computer-aided disease monitoring system is employed for the detection of anomalous and further identification of the abnormalities. The morphological, statistical or localized information differentiate the anomalous from other similar objects or regions [29,30]. Naıve Bayes classifier is used for the classification of diseases from various biomedical signals and clinical images, with low computation burden [31,32]. To obtain higher classification indices using Naıve Bayes classifier, large number of samples is required. K-nearest neighbor (KNN), a supervised learning algorithm, is extensively used in biomedical application for the classification of various diseases [33,34]. KNN classifier, a robust algorithm for noisy training data, has a memory limitation problem. Neural network based classification techniques have been widely explored to classify different biomedical abnormalities with higher degree of accuracy with low computational complexity [35—38]. Support vector machine (SVM), a kernel based discriminative classification model is comprehensively used for the identification of wide varieties of physiological disorders from different biosignals and biomedical images [39—45]. The SVM introduces an optimal separating hyperplane to discriminate linearly separable or nonseparable classes with reduced overfitting problem. Literature suggests the use of other supervised and unsupervised classification algorithms, such as decision tree [46], random forest [31], Gaussian mixture model [47] and so on in various biomedical applications. Recent development in machine learning techniques has introduced convolutional neural network based approaches for the automatic identification of diseases [48—57].

1.2 Skin abnormalities and noninvasive methods for their detection

Among different types of physiological disorders, skin abnormalities are the chosen field of study for this book. Skin is the largest organ of the human body. For a long time, dermatologists believed the visual examination to be sufficient to diagnose the disease, as it is mostly accessible to the eye. However extensive research has exposed that the most demarcating pathogenesis of skin diseases are invisible to naked eye, hidden beneath the skin surface. Moreover, the visible signs of disorder appear at matured stage of the disease, without any insight of the pathogenesis [58]. As a result, lack of detailed observation may lead to improper diagnosis at an early stage of the disease. The incidence of skin cancer has been increasing rapidly all over the world and early treatment is becoming a major issue. Development in bioengineering and medical imaging techniques have introduced ingenious, noninvasive and noncontact devices for in-depth visualization of the pigmentation. Dermatologists follow different standard rules of dermoscopy for the proper identification of the disease. Despite world wide effort to standardize the terms related to dermatoscopic findings for making them objective and reproducible, there exists a scope for a CAD system to assist the clinicians and dermatologists in decision making, especially for investigating an outsized number of patients' data in a shorter duration of time. In the subsequent sections, skin anatomy, various diseases and their clinical aspects are discussed, followed by standard diagnostic rules and monitoring techniques.

1.2.1 Skin surface—structure and properties

Skin is the major part of integumentary system, accounting for around 15% of total adult body weight. Skin plays various important roles in human physiological system [59], such as:

- maintaining homeostasis, that is, steady states of internal physical and chemical conditions;
- protecting against extraneous physical, chemical or biological intruders;
- forestalling the excess water loss;
- thermal regulation of human body;
- synthesizing chemical substances as keratin to provide physical barrier against most invasion, melanin to avoid excessive penetration of ultraviolet radiation and vitamin D for the absorption of calcium;
- discharging waste materials such as ammonia, urea and salts.

Structure of the skin varies according to thickness, color, and textural properties. Depending on the structural properties, skin is categorized into two major types, thick and hairless (found on palms and soles of feet) and thin and hairy (over most of the body). Skin is composed of three layers, namely, the epidermis, the dermis and subcutaneous tissue [60].

- *Epidermis*: Epidermis is the outermost thin, stratified, squamous epithelial layer of skin that lacks blood vessels to supply nutrition and blood in the subsequent layers. The epidermis layer consists of two types of cells: keratinocytes and dendritic cells. Majority of the cells in the epidermis are ectodermally derived keratinocytes, synthesize tough, protective keratin protein. According to the morphology of keratinocyte, the epidermis is subdivided into four layers, namely, basal cell layer, squamous cell layer, granular cell layer, and cornified or horny cell layer. Epidermis is a dynamic layer, giving rise to derivative structures like nail and sweat gland. About 8% of the epidermal cells are melanocytes. Melanocyte cells in stratum basale layer yield the protein pigment melanin, which is responsible for skin color and absorption of harmful ultraviolet light [61–63].
- *Dermis*: Dermis, an integrated system, comprises fibrous connective tissues containing arterioles for supplying oxygen, glucose, water and ions to its structures and

epidermis. Principal component of dermis is collagen, a major stress-resistance material for skin. Dermis layer contains arrector pili muscles to wrinkle the skin and erect the hair. This layer contains the nerves and nerve receptors for the detection of sensation. The dermis and epidermis layers interact with each other to maintain the properties of both tissues [61,63].

- *Subcutaneous tissue*: Subcutaneous layer, also known as hypodermis, shields the skin from injury. The subcutaneous tissue provides the body with buoyancy and functions as a source of energy [62].

1.2.2 Taxonomy of skin diseases

Skin abnormalities are the fourth highest nonfatal disease burden, and are considered to be of major concern to the global health. A wide variety of skin disease occur with various skin conditions. The spectrum of skin diseases is commonly categorized into three major classes: (1) benign, (2) malignant, and (3) non-neoplastic lesions.

1.2.2.1 Benign

Benign skin lesions are noncancerous skin growth. They are generally recognized as dermal, epidermal and melanocytic lesions. Benign melanocytic lesions, such as dysplastic nevus can progress to irreversible malignancy. Seborrheic keratoses (SKs) are examples of benign epidermal lesions and lipoma, fibroma are of benign dermal category.

- *Dysplastic nevi*: Dysplastic nevi, also known as atypical mole may occur in any location of the body, specifically areas exposed to the sunlight. The lesions turn up with irregular edges, rough surface, and "pebbly" appearance and sometimes large size (more than 5 mm diameter). Dysplastic nevus appears in mixture of tan, brown or red/pink color. Presence of these clinically significant features make it difficult to the doctors and radiologist to properly interpret the disease at an early stage [64,65].
- *Seborrheic keratosis*: SK is a noncancerous skin growth and can be found anywhere on the body except on the soles of the feet or palms. SKs are usually brown in color and round or oval in shape, and may sometimes be misinterpreted as suspicious lesion [66,67].

Sample images of dysplastic nevi and SK are shown in Fig. 1.2.

1.2.2.2 Malignant

Malignant lesions are common skin abnormalities and often exhibit assimilated precursor conditions. The lesions develop on exposed skin areas and grow and spread rapidly and often resemble malignancy. These lesions are of primary concern to the physicians for early diagnosis. For the presumptive diagnosis of the disease, dermatologist or expert considers the patient's risk factor, family history, location of the lesion,

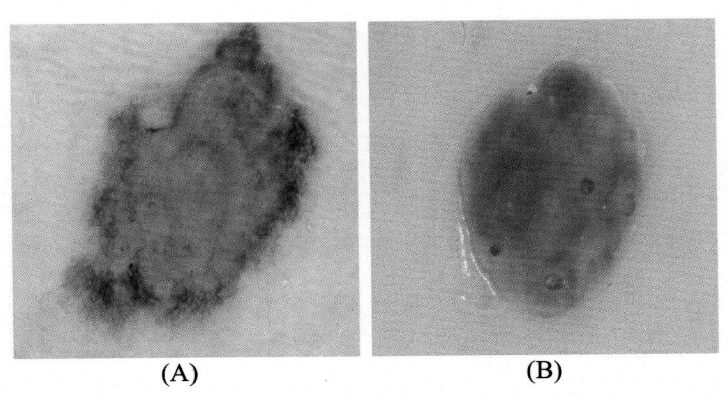

Figure 1.2 Sample images of (A) dysplastic nevi and (B) seborrheic keratosis [68]. *D. Gutman, et al. Skin lesion analysis toward melanoma detection: a challenge at the international symposium on biomedical imaging (ISBI) 2016, hosted by the international skin imaging collaboration (ISIC), 2016 [Online]. Available: <https://arxiv.org/abs/1605.01397>.*

morphological features. To make conclusive remark, doctors rely on the result of the histopathological exam or biopsy of the lesion. Some of the major skin cancers are discussed here.

- *Basal cell carcinoma (BCC)*: BCC is a non-melanoma skin cancer, comprising 60% of primary skin cancers. This type of skin cancer affects the stratum basal cells and terminates the formation of keratin and eventually despoils the dermis. Usually, BCC is a slow growing lesion that despoils the tissues but rarely spreads to the other parts of the body. BCCs commonly occur in face or other uncovered skin surfaces with a small round or oval shaped skin thickening. The lesions usually grow with irregular edges and are uneven in shape, creating atrophic central area and often contain ulcerated vessels. Other rare forms of BCC include superficial basal cell carcinoma, pigmented basal cell carcinoma, infiltrating basal cell carcinoma and so on [69–71].
- *Squamous cell carcinoma (SCC)*: SCC is another type of non-melanoma skin cancer, arising from keratinocytes in stratum spinosum. SCCs typically develop on sun-exposed areas of the skin, for example, head or back of the hands in the elderly. The affected area became reddish, scaling, crevasse with an uneven surface containing visible superficial dilated vessels. SCCs may spread laterally from uneven edges with cluster of lesions. Other skin lesions closely related to SCCs are keratoacanthoma and verrucous carcinoma [71–74].
- *Malignant melanoma*: Malignant melanoma is rare but potentially fatal skin cancer of melanocyets in stratum basal. Malignant melanoma is the fastest growing skin cancer and causes majority of deaths all over the world, predominantly in fair skinned population living in sunny climates. Unlike other skin cancers, malignant melanoma occur in any skin locations irrespective of the regions exposed to excessive

of sunlight [75]. There are four types of malignant melanoma encountered in dermatology. Dark brown or black color superficial melanoma with irregular edges, spread slowly. This is the commonest type of melanoma [76,77]. In comparison with superficial melanoma, the nodular melanoma grows vertically and is more aggressive in nature, developing more abruptly, being noticeable in shiny black dome [78]. Lentigo maligna melanoma seems to be a large flat lesion, having lower risk of metastasis compared to the other types [79]. Darker-skinned patients are most likely to be affected by acral lentiginous melanoma, considered to be one of the rarest skin cancer, appearing in large size with average diameter of 3 cm. It is extremely intrusive and usually develops on the palms, soles of the feet or beneath the nail beds. Patients with malignant melanoma have the risk factors like the presence of dysplastic moles, personal history of melanoma, family history of melanoma or severe blistering sunburn. The clinical features like asymmetric shape with irregular outline, uneven shades of color, and relatively larger size lead to the diagnosis of malignant melanocytic lesions [77]. Since the pigmentation of all the malignant melanoma lesions are not visible through naked eye examination, sometimes the lesions with abnormal growth or bleeding are considered precursors of malignant melanoma. Often atypical mole or dysplastic nevi may turn into melanoma. The presence of 10–12 dysplastic nevi increases the risk of developing malignant melanoma. At the early stage, suspicious features, like location, size, irregular structures, and presence of different significant colors of lesion make it difficult to identify the disease. Blue nevi may point toward melanocytic malignant change [80–83]. Identification at the primitive stage of such lesions is important, as complete removal will cure all type of skin cancer in preliminary stage.

Sample images of BCC, SCC, and malignant melanoma are shown in Fig. 1.3.

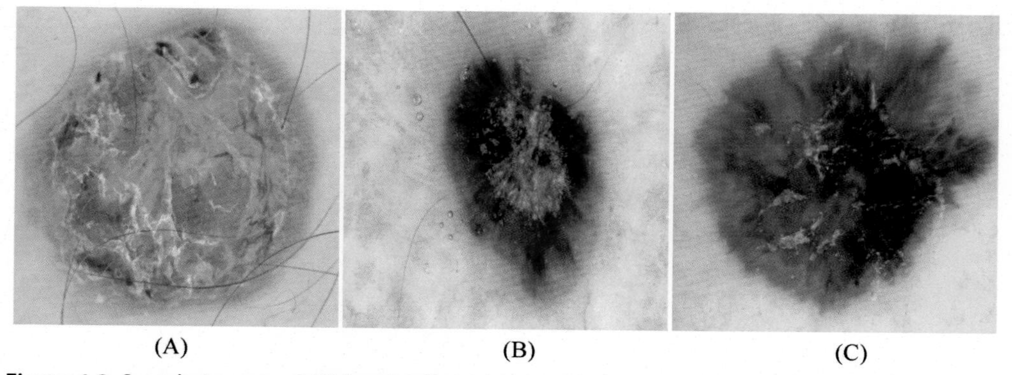

Figure 1.3 Sample images of (A) basal cell carcinoma, (B) squamous cell carcinoma, and (C) malignant melanoma. *D. Gutman, et al. Skin lesion analysis toward melanoma detection: a challenge at the international symposium on biomedical imaging (ISBI) 2016, hosted by the international skin imaging collaboration (ISIC), 2016 [Online]. Available: <https://arxiv.org/abs/1605.01397>.*

1.2.2.3 Nonneoplastic

Wide spectrum of skin diseases belongs to non-neoplastic category. Usually nonneoplastic lesions are found in diverse forms, but many cases have similar clinical appearance. Due to this restricted clinical presentation such as hyperpigmentation, hypopigmentation, macules, nodules and so on, differentiation of each of these different class lesions of utmost importance for further treatment and prediction of prognosis of the disease. However, nonneoplastic skin lesions are easily distinguishable from neoplastic or melanocytic lesions. Some common nonneoplastic skin abnormalities are discussed below:

- *Acne*: Abnormalities associated with sebaceous gland is categorized as acne. Among wide varieties of acnes, acne vulgaris affects the teenagers and acne rosacea typically affects adults. This type of disease occur when hair follicles are clogged with dead skin cells [84,85].
- *Psoriasis*: Psoriasis is an inflammatory skin disease, categorized as reddened epidermal lesion covered by dry silvery scales. Early diagnosis and treatment help to cure such chronic disease [86].

Examples of some nonneoplastic images as acne and psoriasis are shown in Fig. 1.4.

1.2.3 Noninvasive skin imaging techniques

Skin imaging comprises of various visual non-invasive skin analysis techniques. Advanced digital imaging has surpassed the capabilities of human eye and visual cortex and facilitates reliable storage and further monitoring of the information. The objective of skin imaging is not only to study normal living skin, but also includes the pathological, monitoring and diagnostic requirements. The pathological processes require to examine the inflammation and thickness variation of the affected skin area followed by non-invasive monitoring of the disease [87]. For the pigmented skin lesions, the skin

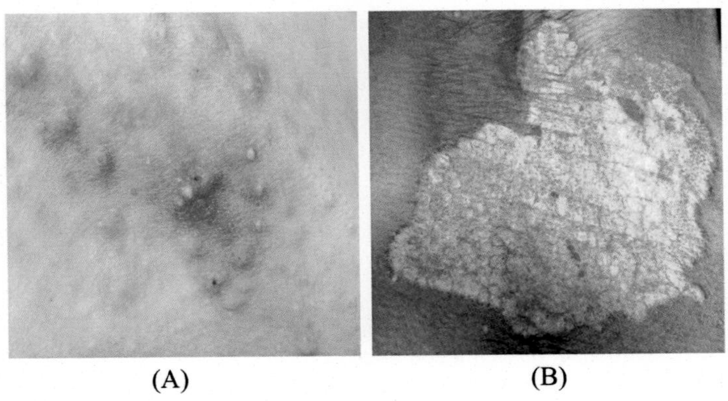

Figure 1.4 Sample images of (A) acne and (B) psoriasis.

imaging devices act as a diagnostic aid. Different skin surface imaging techniques are discussed here with their potential usefulness to the dermatologists.

1.2.3.1 Photography

Digital photography is one of the broadly used scheme of skin imaging. Polarized or ultraviolet light in visible spectrum of 450–700 nm is used to assess the morphological properties of skin surface, like skin lines and wrinkles. Introduction of the polarizing filter brings out more in-depth features from erythema and pigmented lesions. Digital photography has a great impact on condition monitoring of skin abnormalities and teledermatology [87,88].

1.2.3.2 Profilometry

Profilometry is a leading skin surface topography assessment technique. Optical profilometry is a widely used technique for analysis of depths of skin surface defects. Laser profilometry using dynamic focusing is a very useful technique for the evaluation of skin line disruption [87,88].

1.2.3.3 Optical coherence tomography

OCT, popularly known for retinal imaging is used to obtain the descriptive morphology of skin surface. It uses low coherence laser in near infrared range (700–2500 nm) to study the tissue morphology in subsurface level (depth of penetration 1–2 mm) [87,88].

1.2.3.4 High-resolution ultrasonography

High-resolution ultrasonography, using very high frequency sound wave (13.5–100 MHz) is confined to the varying acoustic impedance of tissues. It is a very useful diagnostic tool for the assessment of skin tumors, monitoring of inflammatory disorders, preoperative tumor depth estimation, and so on [87,88].

From a wide spectrum of skin imaging techniques, epiluminescence microscopy or dermoscopy has been considered in the present book for the assessment and condition monitoring of skin diseases. A detailed overview of dermoscopic imaging and its role in skin disease identification is given in the subsequent sections.

1.3 An overview of dermoscopy

Dermoscopy or dermatoscopy or epiluminescence microscopy is a noninvasive, micromorphological imaging technique, used for in-detailed visualization of important features of pigmented skin lesions. Differentiation of a surprisingly large number of skin lesions, between melanocytic and nonmelanocytic or melanoma and nonmelanoma or benign and malignant is difficult in naked eye examination. Dermoscopy is a standard technique employed for microscopic morphological investigation of skin lesions. This epiluminescence microscopy provides a magnification range from 60× to 80× for the screening of skin cancer [77]. The recent development of epiluminescence microscopy employs an additional optical system for simultaneous view and a camera for instant photography. This epiluminescence microscopic system equipped with digital image acquisition module is referred to as digital epiluminescence dermatoscope [75–78].

1.3.1 Types of dermoscopy

Depending on integrated transilluminating light source and associated optical magnification, dermoscopy is categorized in two different modes, namely (1) nonpolarized dermoscopy and (2) polarized dermoscopy.

1.3.1.1 Nonpolarized dermoscopy

In nonpolarized dermoscope (NPD), a magnifying lens and a light emitting diode (LED) as light source are integrated to visualize the subsurface morphology of cutaneous lesions beneath the superficial dermis layer. NPD makes a direct interface with the skin surface through a glass plate and immersion liquid as ultrasound gel, 70% alcohol or mineral oil. The air bubbles at the interface of dermoscope's glass plate and immersion liquid introduce artifacts in the acquired dermoscopic image [87]. In NPD, superficial penetrating light encounters minimal scattering events and is considered to be the source of light that is penetrated or reflected back at the epidermis layers and the dermal–epidermal junction. Thus NPD allows the interpretation of subsurface structures of epidermis and dermal–epidermal junction. However, it is unable to extract the structures from deeper layers due to small fractional contribution of deep penetrating light in the back reflected light. The dermoscopic images acquired using NPD explore the color and morphological structures, not visible to the unaided eye. The detailed visualization of pigmented and nonpigmented lesions improves the diagnostic performance of the expert dermatologists [88–93].

1.3.1.2 Polarized dermoscopy

Similar to NPD, polarized dermoscope uses magnifying lens along with LED as source of illumination. Two polarizing filters are employed in polarized dermoscope to realize cross-polarization of light that ensures the acquisition of image without direct contact to the skin and use of immersion liquid. In polarized dermoscopy, polarized light is generated by passing the light from LED through a polarizer and subsequently the reflected light is sensed by the detector through a cross-polarized filter. However, the polarized light can pass through the cross-polarizing filter until the original polarized light has sufficient scattering in the skin. The light reflected from the stratum corneum and absorbed in superficial layer of epidermis does not encounter sufficient scattering to implicate randomization of polarization [77,91]. However, randomization of polarization is obtained only when light penetrates more deeply and undergo multiple scattering. In polarized dermoscopy, deep penetrating light is acquired by cross-polarization filter due to randomization of polarization and allow the detailed visualization of dermoscopic structures from superficial dermis and dermal–epidermal junction [88–93].

1.3.2 Role of dermoscopy in skin lesion identification

Dermoscopy or dermatoscopy yields information about various skin pigmentations with different magnifications that lead to considerably improved differential diagnosis of the disease from the additional submicroscopic information. Dermoscopy is a standard tool for the early-stage detection and monitoring of malignant melanocytic lesions, which sometimes become very much challenging even to the experienced dermatologists. Dermoscopy is extensively used in dermatology to extract the dominant criteria for diagnosis, like (1) morphology of vascular structure, (2) texture pattern, (3) colors, (4) follicular abnormalities, and (5) specific dermoscopic features from pigmented skin lesions [87,93]. The detailed visualization of dermoscopic features, not accessible through visual inspection, assists the radiologists or expert dermatologists to differentiate skin abnormalities without unnecessary invasive histopathological examination at an early stage.

Examples of some handheld dermoscopes available in the marketare: Delta 20T (Heine), Dermlite (3Gen), Veos (Canfield). Several companies provide dermoscopes compatible with smart phone or tablets such as Dermlite (3Gen), Handyscope (Fotofinder), Heine R-IC1 and so on, for image acquisition, storage and transfer to the medical personnel [94].

1.4 Standard dermoscopic rules for skin diseases diagnosis

Dermoscopy or epiluminescence microscopy uncovers the detailed morphological structures and visual properties of pigmented skin lesions. This detailed visualization of dermoscopic findings has improved the diagnostic accuracy by 5%–30%. Incorporating this lesion screening tool, the expert dermatologists use some simplified rule-based clinical algorithms such as ABCD rule of dermoscopy [95], seven-point checklist [96,97], Menzies method [98] and so on for the diagnosis of skin diseases. Among these dermoscopic rules for skin disease identification, ABCD rule of dermoscopy is most widely explored by the expert dermatologists. In this book, the dermoscopic features related to ABCD rule of dermoscopy are considered to develop an indigenous integrated system for skin disease identification and monitoring. ABCD rule and its associated dermoscopic features are elaborated in the subsequent stages.

1.4.1 ABCD rule of dermoscopy

The ABCD rule of dermoscopy was first introduced to differentiate between benign and malignant skin lesions. Here, the ABCD attributes are developed to quantify the dermoscopic findings and subsequently categorize the lesions in benign, suspicious and malignant classes. The ABCD rule of dermoscopy has incorporated the clinical criterion of skin lesions as asymmetry (A), border irregularity (B), dermoscopic colors (C) and differential structures (D) [95,99]. Quantifying these clinical features, a scoring system is developed to evaluate total dermoscopic score (TDS) to compartmentalize the lesions in benign, suspicious and malignant category.

- *Asymmetry (A)*: The structural variations, in terms of the contour, dermoscopic color and differential structures along the skin lesion area are considered to evaluate the asymmetry attribute of the ABCD rule. To quantify the degree of asymmetry, the skin lesion is bisected into symmetric planes along two 90° axes and asymmetry is evaluated by considering the color and structural distribution along either side of each axis. To estimate the asymmetry of the lesion, an asymmetricity score has been assigned from the minimum value of 0 to the maximum value of 2. If asymmetry is absent with respect to both the axes, then the asymmetry score is zero. Similarly, a score of one is assigned for single axis asymmetry and two for both axes asymmetry. To calculate the asymmetry score, "A" attribute is multiplied by a constant value of 1.3 [95,99].
- *Border (B)*: The border irregularity measure estimates the irregularity or abruptness of pattern at the periphery of the lesion. To evaluate the border irregularity score, the border of the skin lesion is divided into eight quadrants. Maximum score of eight is assigned when the entire lesion border (all the segments) has a distinct cut off. Similarly, minimum score of zero is assigned for indistinct cut off at the entire lesion periphery. To calculate the "B" score, the minimum value of zero to maximum value of eight is multiplied by a constant 0.1 [95,99].
- *Color (C)*: The significant dermoscopic colors like light and dark brown indicate the distribution of melanin in epidermis or superficial dermis layer. Black color reveals the presence of melanin in upper granular or stratum corneum layer and blue gray reflects the melanin in papillary dermis layer. Similarly, neovascularization is revealed by the presence of red color, while white color corresponds to regression area. The "C" score is estimated based on the presence of these six clinically significant colors for the differentiation of skin lesions. Minimum score of one to maximum of six is considered for the presence of each of the six colors. A constant value of 0.5 is multiplied by the "C" score to estimate the C attribute of ABCD rule [95,99].
- *Differential structures (D)*: For the characterization of skin lesion, presence of differential structures is considered to be an important dermoscopic finding. To determine

the dermoscopic structural features, presence of structureless area, pigment network, branched streaks, dots and globules have been given the major consideration. To estimate the dermoscopic score for differential structures, a maximum of five score is allotted, one score to each structure [95,99].

After estimating the A, B, C and D attributes, the skin lesions are classified based on TDS. The TDS score is estimated according to the following equation [95,99].

$$TDS = A \times 1.3 + B \times 0.1 + C \times 0.5 + D \times 0.5$$

According to the standard ABCD rule of dermoscopy, low TDS ($TDS < 4.75$) determines the benign lesions, while the intermediate score between 4.75 and 5.45 corresponds to suspicious lesions. The high score of TDS indicates malignant lesions [95,99].

1.5 Computer-aided skin disease monitoring system

The unaided clinical examination of pigmented skin lesions, which has limited and variable identification accuracy, leads to important challenges in early detection of the disease and consequent minimization of unnecessary biopsy. Visual similarity in the manifestation of different skin lesions in terms of morphological, textural and color complexity makes early diagnosis not an easy task for both the general physicians and dermatologists. Development of high-resolution dermoscopic imaging technique allows the dermatologists and radiological experts to visualize the detailed skin structures and significant dermoscopic findings. Despite world wide effort to standardize the terms of dermatoscopic findings so that such findings are not subjective and they become reproducible, there exists a scope for a CAD system to assist the clinicians and dermatologists in decision making, especially in the cases of investigating an outsized number of patients' data in a shorter duration of time. The development of a CAD system plays an important role in accurate and uniform evaluation of skin disorders at an early stage, for further medication and prevention of the disease. The CAD is an important tool for the quantification of dermoscopic findings and act as a second opinion apart from the expert's opinion for the diagnosis of the disease. The basic framework of a typical CAD system for skin disease monitoring comprises image preprocessing module, segmentation, feature extraction and feature selection module followed by a classification module. A schematic representation of a computer-based skin disease diagnostic system is shown in Fig. 1.5.

The acquired images of skin lesions may contain noise due to inexact illumination. Depending on the anatomical sites, the dermoscopic images are contaminated with thick and thin hair artifacts. The image preprocessing algorithm serves to eliminate the noise and artifacts from the input image to improve the image quality and further ensure the

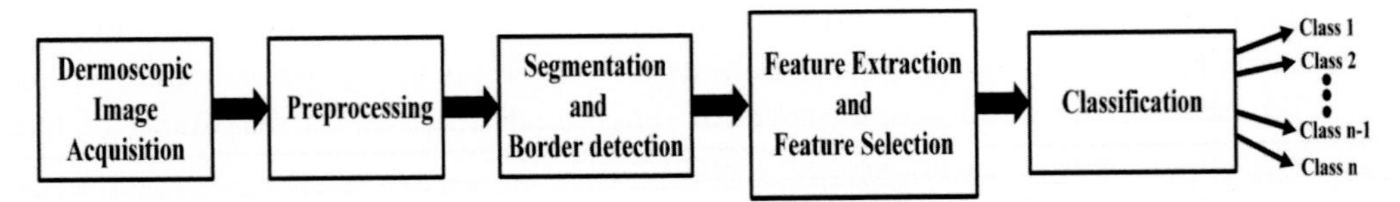

Figure 1.5 Block diagram representation of a typical computer-aided skin disease monitoring system.

development of efficient algorithms in subsequent stages. Image segmentation technique segregates the pigmented lesion from the surrounding normal skin region. The proper segmentation of the lesion area helps to identify the morphological properties and its further spreading throughout the affected area. Effective segmentation of the region of interest leads to the detection of the border to detect the irregularity of the lesion. Feature extraction module implementing advanced signal processing tools, quantifies the morphological properties and clinically significant dermoscopic criteria. Feature extraction algorithm extracts the statistical information and clinically correlated features to differentiate visually similar pigmented lesions. From the extracted features, the feature selection algorithm identifies the most demarcating features for further classification of the disease. The classification module is responsible for identifying the skin abnormalities from the significant features, with a higher degree of accuracy.

In the last few decades, an extensive research has been carried out on dermoscopic image analysis, to identify various skin abnormalities with improved performance indices. Barata et al. have proposed two systems for melanoma identification, using global and local textural and color features with 96% sensitivity and 80% specificity [100]. Simizu et al. have introduced a task decomposition based four-class classification technique to differentiate melanocytic and nonmelanocytic skin lesions. Here, melanoma and nevus are considered in melanocytic while BCCs and SKs are in nonmelanocytic lesion category. The proposed scheme has identified melanoma, nevus, BCC and SK with 90.48%, 82.51%, 82.61%, and 80.61% identification rate [101]. Rastgoo et al. have differentiated melanoma and dysplastic nevi, considering the combined feature set of texture, shape and color [102]. Abuzaghleh et al. have reported a noninvasive real-time automated skin lesion analysis system for the differentiation of benign, atypical and melanoma. The proposed scheme has used spatial and spectral features along with color and structural information to classify the skin diseases [103]. Garnavi et al. have proposed a wavelet based texture analysis technique and boundary-series analysis for the diagnosis of melanoma with an accuracy of 91.26% [104]. Jiji and DuraiRaj have proposed a content based image retrieval technique for the extraction of visual information from the dermoscopic images, and particle swarm optimization technique for multiclass classification of skin lesions [105].

1.6 Scope of the book

In this book, an automatic skin disease identification technique using digital signal processing tools has been described. A set of simple and efficient algorithms for preprocessing, segmentation, feature extraction and classification of the diseases from dermoscopic images have been illustrated. The organization of the book is as follows:

Chapter 2

This chapter will address the preprocessing techniques and their importance for dermoscopic images followed by segmentation of skin lesions.

Chapter 3

Development of hand crafted image processing tools for morphological, texture feature extraction from skin lesion images will be discussed here.

Chapter 4

State-of-the-art techniques for selection and learning of appropriate and demarcating features followed by binary and multiclass classification techniques will be demonstrated in this chapter.

Chapter 5

In this chapter, step-by-step method for developing an integrated expert system implementing expert's knowledge to replicate gold standard rule of dermoscopy will be elaborated.

Chapter 6

Finally, in this chapter, the book will be concluded with ideas of future scope of research in computer-aided diagnosis of biomedical systems using digital signal processing and machine learning tools.

References

[1] T. Newman, A brief introduction to physiology, Medical News Today, MediLexicon International Ltd., 13 October 2017 (Web 27.08.19).

[2] R.M. Rangayyan, "Biomedical Signal Analysis, John Wiley and Sons, 2002.

[3] M. Chan, E. Campo, D. Brulin, D. Estève, Biomedical monitoring technologies and future healthcare systems, J. Sci. Tech. 3 (1) (2017) 54–79.

[4] A. Astaras, P.D. Bamidis, C. Kourtidou-Papadeli, N. Maglaveras, Biomedical real-time monitoring in restricted and safety-critical environments, Hippokratia 12 (2008) 10–14.

[5] J.L. Semmlow, Biosignal and Biomedical Image Processing, Marcel Dekker, 2004.

[6] J.G. Webster, Medical Instrumentation Application and Design, fourth ed., John Wiley & Sons, 2010.

[7] J.G. Webster, second ed., Encyclopaedia of Medical Devices and Instrumentation, 5, John Wiley & Sons, 2006.

[8] J.G. Fujimoto, C. Pitris, S.A. Boppart, M.E. Brezinski, Optical coherece tomography: an emerging technology for biomedical imaging and optical biopsy, Neoplasia 2 (1–2) (2000) 9–25.

[9] Y. Lin, Y.H. Hu, Power line interference detection and suppression in ECG signal processing, IEEE Trans. Biomed. Eng 55 (1) (2008) 354–357.

[10] J. Oster, J. Behar, O. Sayadi, S. Nemati, A.E.W. Johnson, G.D. Clifford, Semisupervised ECG ventricular beat classification with novelty detection based on switching Kalman filters, IEEE Trans. Biomed. Eng. 62 (9) (2015) 2125–2134.
[11] C. Orphanidou, I. Drobnjak, Quality assessment of ambulatory ECG using wavelet entropy of the HRV signal, IEEE J. Biomed. Health Inform. 21 (5) (2017) 1216–1223.
[12] U.R. Acharya, Y. Hagiwara, S.N. Deshpande, S. Suren, J.E.W. Koh, S.L. Oh, et al., Characterization of focal EEG signals: review, Future Gen. Comp. Sysm. 91 (2019) 290–299.
[13] A.G. Ramakrishnan, A.P. Prathosh, T.V. Ananthapadmanabha, Threshold-independent QRS detection using the dynamic plosion index, IEEE Signal Proc. Lett 21 (5) (2014) 554–558.
[14] C. Nayak, S.K. Saha, R. Kar, D. Mandal, An efficient and robust digital fractional order differentiator based ECG pre-processor design for QRS detection, IEEE Trans. Biomed. Circ. Syst 13 (4) (2019) 682–696.
[15] H.M. Rai, K. Chatterjee, Hybrid adaptive algorithm based on wavelet transform and independent component analysis for denoising of MRI images, Measurement 144 (2019) 72–82.
[16] S. Pal, S. Chatterjee, D. Dey, S. Munshi, Morphological operations with iterative rotation of structuring elements for segmentation of retinal vessel structures, Multidim Syst. Signal Process. 30 (2019) 373–389.
[17] S. Sun, Y. Guo, Y. Guan, H. Ren, L. Fan, Y. Kang, Juxta-vascular nodule segmentation based on flow entropy and geodesic distance, IEEE J. Biomed. Health Inform. 18 (4) (2014) 1355–1362.
[18] K.M. Adal, P.G. van Etten, J.P. Martinez, K.W. Rouwen, K.A. Vermeer, L.J. van Vliet, An automated system for the detection and classification of retinal changes due to red lesions in longitudinal fundus images, IEEE Trans. Biomed. Eng. 65 (6) (2018) 1382–1390.
[19] S. Mitra, C. Balaji, A neural network based estimation of tumor parameters from a breast thermogram, Int. J. Heat Mass Transf. 53 (2010) 4714–4727.
[20] B.L. Koley, D. Dey, Real-time adaptive apnea and hypopnea event detection methodology for portable sleep apnea monitoring devices, IEEE Trans. Biomed. Eng. 60 (12) (2013) 3354–3363.
[21] T. Mar, S. Zaunseder, J.P. Martínez, M. Llamedo, R. Poll, Optimization of ECG classification by means of feature selection, IEEE Trans. Biomed. Eng. 58 (8) (2011) 2168–2177.
[22] C.C. Lin, C.M. Yang, Heartbeat classification using normalized RR intervals and morphological features, Math. Probl. Eng. 2014 (2014) (Art. no. 712474).
[23] S. Mohamed, S. Haggag, S. Nahavandi, O. Haggag, Towards automated quality assessment measure for EEG signals, Neurocomputing 237 (2017) 281–290.
[24] X. Lin, H. Fan, H. Wang, L. Wang, Common spatial patterns combined with phase synchronization information for classification of EEG signals, Biomed. Signal Procs Control. 52 (2019) 248–256.
[25] P. Sahu, D. Yu, M. Dasari, F. Hou, H. Qin, A lightweight multi-section CNN for lung nodule classification and malignancy estimation, IEEE J. Biomed. Health Inform. 23 (3) (2019) 960–968.
[26] C. Barata, M.E. Celebi, J.S. Marques, A survey of feature extraction in dermoscopy image analysis of skin cancer, IEEE J. Biomed. Health Inform. 23 (3) (2019) 1096–1109.
[27] J. Banerjee, Y. Sun, C. Klink, R. Gahrmann, W.J. Niessen, A. Moelker, et al., Multiple-correlation similarity for block-matching based fast CT to ultrasound registration in liver interventions, Med. Image Anal. 53 (2019) 132–141.
[28] K. Ram, G.D. Joshi, J. Sivaswamy, A successive clutter-rejection-based approach for early detection of diabetic retinopathy, IEEE Trans. Biomed. Eng. 58 (3) (2011) 664–673.
[29] Y. Masutani, H. MacMahon, K. Doi, Computerized detection of pulmonary embolism in spiral CT angiography based on volumetric image analysis, IEEE Trans. Med. Imaging 21 (12) (2002) 1517–1523.
[30] I. Sluimer, A. Schilham, M. Prokop, B. van Ginneken, Computer analysis of computed tomography scans of the lung: a survey, IEEE Trans. Med. Imaging 25 (4) (2006) 385–405.
[31] L. Atallah, B. Lo, R. King, G.-Z. Yang, Sensor positioning for activity recognition using wearable accelerometers, IEEE Trans. Biomed. Circ. Syst. 5 (4) (2011) 320–329.

[32] D. Jain, V. Singh, Feature selection and classification systems for chronic disease prediction: a review, Egypt Inform. J. 19 (2018) 179–189.
[33] P.M. Bentley, P.M. Grant, J.T.E. McDonnell, Time-frequency and time-scale techniques for the classification of native and bioprosthetic heart valve sounds, IEEE Trans. Biomed. Eng. 45 (1) (1998) 125–128.
[34] A. Sengur, I. Turkoglu, A hybrid method based on artificial immune system and fuzzy k-NN algorithm for diagnosis of heart valve diseases, Expert Syst. Appl. 35 (3) (2008) 1011–1020.
[35] A. Moukadem, A. Dieterlen, N. Hueber, C. Brandt, A robust heart sounds segmentation module based on S-transform, Biomed. Signal Process. Control. 8 (3) (2013) 273–281.
[36] C.N. Gupta, R. Palaniappan, S. Swaminathan, S.M. Krishnan, Neural network classification of homomorphic segmented heart sounds, Appl. Soft Comput. 7 (1) (2007) 286–297.
[37] H. Uguz, A biomedical system based on artificial neural network and principal component analysis for diagnosis of the heart valve diseases, J. Med. Syst. 36 (1) (2012) 61–72.
[38] S. Kang, R. Doroshow, J. McConnaughey, R. Shekhar, Automated identification of innocent Still's murmur in children, IEEE Trans. Biomed. Eng. 64 (6) (2017) 1326–1334.
[39] W. Zhang, J. Han, S. Deng, Heart sound classification based on scaled spectrogram and tensor decomposition, Expert Syst. Appl. 84 (2017) 220–231.
[40] S. Sun, An innovative intelligent system based on automatic diagnostic feature extraction for diagnosing heart diseases, Knowl. Based Syst. 75 (2015) 224–238.
[41] F. Safara, S. Doraisamy, A. Azman, A. Jantan, A.R.A. Ramaiah, Multi-level basis selection of wavelet packet decomposition tree for heart sound classification, Comput. Biol. Med. 43 (10) (2013) 1407–1414.
[42] Y. Zheng, X. Guo, J. Qin, S. Xiao, Computer-assisted diagnosis for chronic heart failure by the analysis of their cardiac reserve and heart sound characteristics, Comput. Methods Prog. Biomed. 122 (3) (2015) 372–383.
[43] W. Zhang, J. Han, S. Deng, Heart sound classification based on scaled spectrogram and partial least squares regression, Biomed. Signal Process. Control 32 (2017) 20–28.
[44] Y. Zheng, X. Guo, X. Ding, A novel hybrid energy fraction and entropy-based approach for systolic heart murmurs identification, Expert Syst. Appl. 42 (5) (2015) 2710–2721.
[45] B.M. Whitaker, P.B. Suresha, C. Liu, G. Clifford, D. Anderson, Combining sparse coding and time-domain features for heart sound classification, Physiol. Meas. 38 (2017) 1701–1713.
[46] R.C. King, E. Villenueve, R.J. White, R.S. Sherratt, W. Holderbaum, W.S. Harwin, Application of data fusion techniques and technologies for wearable health monitoring, Med. Eng. Phys. 42 (2017) 1–12.
[47] A.M. Mannini, A. Sabatini, Machine learning methods for classifying human physical activity from on-body accelerometers, Sensors 10 (2) (2010) 1154–1175.
[48] H. Shin, H.R. Roth, M. Gao, L. Lu, Z. Xu, I. Nogues, et al., Deep convolutional neural networks for computer aided detection: CNN architectures, dataset characteristics and transfer learning, IEEE Trans. Med. Imaging 35 (5) (2016) 1285–1298.
[49] A. Krizhevsky, Learning Multiple Layers of Features From Tiny Images (M.S. thesis), Department of Computer Science, University of Toronto, Toronto, Canada, 2009.
[50] A. Krizhevsky, I. Sutskever, G.E. Hinton, ImageNet classification with deep convolutional neural networks, in: Proceedings of the NIPS, pp. 1097–1105, 2012.
[51] C. Szegedy, et al., Going deeper with convolutions, in: Proceedings of the IEEE Computer Society Conference on Computer Vision and Pattern Recognition, pp. 1–9, 2015.
[52] Y. Song, W. Cai, Y. Zhou, D.D. Feng, Feature-based image patch approximation for lung tissue classification, IEEE Trans. Med. Imaging 32 (4) (2013) 797–808.
[53] M. Gao, U. Bagci, L. Lu, A. Wu, M. Buty, H. Shin, et al., Holistic classification of CT attenuation patterns for interstitial lung diseases via deep convolutional neural networks, in: Proceedings of the MICCAI First Workshop Deep Learning in Medical Image Analysis, 2015.
[54] S.C. Bunce, M. Izzetoglu, K. Izzetoglu, B. Onaral, K. Pourrezaei, Functional near-infrared spectroscopy, IEEE Eng. Med. Biol. Mag. 25 (4) (2006) 54–62.

[55] X. Zhai, B. Jelfs, R.H.M. Chan, C. Tin, Self-recalibrating surface EMG pattern recognition for neuroprosthesis control based on convolutional neural network, Front. Neurosci. 11 (2017) 379. Available from: https://doi.org/10.3389/fnins.2017.00379. Published 11.07.17.
[56] P. Xia, J. Hu, Y. Peng, EMG-based estimation of limb movement using deep learning with recurrent convolutional neural networks, Artif. Organs 42 (5) (2018) E67–E77.
[57] D. Rav'ß, C. Wong, F. Deligianni, M. Berthelot, J. Andreu-Perez, B. Lo, et al., Deep learning for health informatics, IEEE J. Biomed. Health Inform. 21 (1) (2017) 4–21.
[58] A.M. Kligman, The invisible dermatoses, Arch. Dermatol. 127 (1991) 1375.
[59] Skin Anatomy, Physiology, and Assessment, AMN Healthcare in association with Interact Medical, 2014.
[60] P.A.J. Kolarsick, M.A. Kolarsickand, C. Goodwin, Anatomy and physiology of the skin, J. Dermatol. Nurses' Assoc. 3 (4) (2011) 203–213.
[61] D.H. Chu, Overview of Biology, Development, and Structure of Skin (Chapter 7), 2008.
[62] K. Wolff, L.A. Goldsmith, S.I. Katz, B.A. Gilchrest, A.S. Paller, D.J. Leffell (Eds.), Fitzpatrick's Dermatology in General Medicine, seventh ed., McGraw-Hill, New York, 2008, pp. 57–73.
[63] W.D. James, T.G. Berger, D.M. Elston, Andrews' Diseases of the Skin: Clinical Dermatology, tenth ed., Elsevier Saunders, Philadelphia, 2006.
[64] C. Bevona, W. Goggins, T. Quinn, J. Fullerton, H. Tsao, Cutaneous melanomas associated with nevi, Arch. Dermatol. 139 (12) (2003) 1620–1624.
[65] A. Roesch, W. Burgdorf, W. Stolz, M. Landthaler, T. Vogt, Dermatoscopy of 'dysplastic nevi': a beacon in diagnostic darkness, Eur. J. Dermatol. 16 (2006) 479–493.
[66] A.J. Birnie, S. Varma, A dermatoscopically diagnosed collision tumour: malignant melanoma arising within a seborrhoeic keratosis, Clin. Exp. Dermatol. 33 (2008) 512–513.
[67] P. Zaballos, S. Blazquez, S. Puig, E. Salsench, J. Rodero, J.M. Vives, et al., Dermoscopic pattern of intermediate stage in seborrhoeic keratosis regressing to lichenoid keratosis: report of 24 cases, Br. J. Dermatol. 157 (2007) 266–272.
[68] D. Gutman, et al. Skin lesion analysis toward melanoma detection: a challenge at the international symposium on biomedical imaging (ISBI) 2016, hosted by the international skin imaging collaboration (ISIC), 2016 [Online]. Available: <https://arxiv.org/abs/1605.01397>.
[69] J.M. Martín, R. Bella-Navarro, E. Jordá, Vascular patterns in dermoscopy, Actas Dermosifiliogr. (Engl. Ed.) 103 (5) (2012) 357–375.
[70] P. Kharazmi, Md. I. AlJasser, H. Lui, Z.J. Wang, T.K. Lee, Automated detection and segmentation of vascular structures of skin lesions seen in dermoscopy, with an application to basal cell carcinoma classification, IEEE J. Biomed. Health Inform. (2017). Available from: https://doi.org/10.1109/JBHI.2016.2637342.
[71] S.N. Snow, W. Sahl, J.S. Lo, F.E. Mohs, T. Warner, J.A. Dekkinga, et al., Metastatic basal cell carcinoma. Report of five cases, Cancer 73 (1994) 328–335.
[72] C.K. Bichakjian, T. Olencki, S.Z. Aasi, M. Alam, J.S. Andersen, D. Berg, et al., "Basal cell skin cancer, version 1.2016, J. Nat. Comp. Cancer Net. 14 (5) (2016) 574–597.
[73] A.L. Krunic, D.R. Garrod, N.P. Smith, G.S. Orchard, O.B. Cvijetic, Differential expression of desmosomal glycoproteins in keratoacanthoma and squamous cell carcinoma of the skin: an immunohistochemical aid to diagnosis, Acta Derm. Venereol. 76 (1996) 394–398.
[74] E.M. de Villiers, D. Lavergne, K. McLaren, E.C. Benton, Prevailing papillomavirus types in non-melanoma carcinomas of the skin in renal allograft recipients, Int. J. Cancer 73 (1997) 356–361.
[75] R. Marks, An overview of skin cancers: incidence and causation, Cancer Suppl. 75 (S2) (1995) 607–612.
[76] M.E. Vestergaard, P. Macaskill, P.E. Holt, S.W. Menzies, Dermoscopy compared with naked eye examination for the diagnosis of primary melanoma: a *meta*-analysis of studies performed in a clinical setting, Br. J. Dermatol. 159 (2008) 669–676.
[77] H. Pehamberger, M. Binder, A. Steiner, K. Wolff, In vivo epiluminescencemicroscopy: improvement of early diagnosis of melanoma, J. Invest. Dermatol. 100 (1993) 356S–362S.

[78] Menzies, et al., Dermoscopic evaluation of nodular melanoma, JAMA Dermatol. 149 (2013) 699–709.
[79] R. Schiffner, J. Schiffner-Rohe, T. Vogt, M. Landthaler, U. Wlotzke, A.B. Cognetta, et al., Improvement of early recognition of lentigo maligna using dermatoscopy, J. Am. Acad. Dermatol. 42 (2000) 25–32.
[80] T. Saida, et al., Significance of dermoscopic patterns in detecting malignant melanoma on acral volar skin: results of a multicenter study in Japan, Arch. Dermatol. 140 (2004) 1233–1238.
[81] H. Schulz, Epiluminescence microscopy features of cutaneous malignant melanoma metastases, Melanoma Res. 10 (2000) 273–280.
[82] H.C. Williams, January 1 Dermatology, Radcliffe Publishing Ltd, 1997.
[83] P. Das, N. Deshmukh, N. Badore, C. Ghulaxe, P. Patel, A review article on melanoma, J. Pharm. Sci. Res. 8 (2) (2016) 112–117.
[84] E.A. Eady, Topical antibiotics for the treatment of acne, J. Dermatol. Treat. 1 (1990) 215–226.
[85] N.B. Simpson, Social and economic aspects of acne and the cost effectiveness of isotretinoin, J. Dermatol. Treat. 4 (2) (1993) S6–S9.
[86] L.J. Petersen, J.K. Kristensen, Selection of patients for psoriasis clinical trials: a survey of the recent dermatological literature, J. Dermatol. Treat. 3 (1992) 171–176.
[87] J. Serup, G.B.E. Jemec, G.L. Grove, Handbook of Non-Invasive Methods and the Skin, second ed., Taylor & Francis Grp, 2006.
[88] D. Rallan, C.C. Harland, Skin imaging: is it clinically useful? Clin. Dermatol. 29 (2004) 453–459.
[89] R.P. Braun, H.S. Rabinovitz, M. Oliviero, A.W. Kopf, J. Saurat, Dermoscopy of pigmented skin lesions, J. Am. Acad. Dermatol. 52 (2005) 109–121.
[90] H.P. Soyer, G. Argenziano, S. Chimenti, V. Ruocco, Dermoscopy of pigmented skin lesions, Eur. J. Dermatol. 11 (2001) 270–276.
[91] Y. Pan, D.S. Gareau, A. Scope, M. Rajadhyaksha, N.A. Mullani, A.A. Marghoob, Polarized and non-polarized dermoscopy the explanation for the observed differences, Arch. Dermatol. 144 (6) (2008).
[92] A. Lallas, Z. Apalla, G. Chaidemenos, New trends in dermoscopy to minimize the risk of missing melanoma, J. Skin Cancer (2012). Available from: https://doi.org/10.1155/2012/820474.
[93] C.M. Grin, K.P. Friedman, J.M. Grant-Kels, Dermoscopy: a review, Dermatol. Clin. 20 (2002) 641–646.
[94] K.C. Nischal, U. Khopkar, Dermoscope, Indian. J. Dermatol. Venereol. Leprol. 71 (4) (2005) 300–303.
[95] F. Nachbar, et al., The ABCD rule of dermatoscopy: high prospective valie in the diagnosis of doubtful melanocytic skin lesions, J. Am. Acad. Dermatol. 30 (4) (1994) 551–559.
[96] G. Argenziano, G. Fabbrocini, P. Carli, V. De Giorgio, E. Sammarco, M. Delfino, Epiluminescence microscopy for the diagnosis of doubtful melanocytic skin lesions, Arch. Dermatol. 134 (12) (1998) 1563–1570.
[97] J. Kawahara, S. Daneshvar, G. Argenziano, G. Hamarneh, 7-Point checklist and skin lesion classification using multi-task multi-modal neural nets, IEEE J. Biomed. Health Inform. (2018). Available from: https://doi.org/10.1109/JBHI.2018.2824327.
[98] S.W. Menzies, C. Ingvar, K.A. Crotty, W.H. McCarthy, Frequency and morphologic characteristics of invasive melanomas lacking specific surface microscopic features, Arch. Dermatol. 132 (10) (1996) 1178–1182.
[99] A.A. Marghoob, J. Malvehy, R.P. Braun, Atlas of Dermoscopy, second ed., Informa healthcare, 2012.
[100] C. Barata, M. Ruela, M. Francisco, T. Mendonc, J.S. Marques, Two systems for the detection of melanomas in dermoscopic images using texture and color features, IEEE Syst. J. 8 (3) (2014) 965–979.
[101] K. Shimizu, H. Iyatomi, M.E. Celebi, K. Norton, M. Tanaka, Four-class classification of skin lesions with task decomposition strategy, IEEE Trans. Biomed. Eng. 62 (1) (2015) 274–283.
[102] M. Rastgoo, R. Garcia, O. Morel, F. Marzani, Automatic differentiation of melanoma from dysplastic nevi, Comput. Med. Imaging Graph. 43 (2015) 44–52.

[103] O. Abuzaghleh, B.D. Barkana, M. Faezipour, Non-invasive realtime automated skin lesion analysis system for melanoma early detection and prevention, IEEE J. Trans. Eng. Health Med. 3 (2015).
[104] R. Garnavi, M. Aldeen, J. Bailey, Computer-aided diagnosis of melanoma using border and wavelet-based texture analysis, IEEE Trans. Inf. Tech. BioMed. 16 (6) (2012) 1239–1252.
[105] G.W. Jiji, P.J. DuraiRaj, Content-based image retrieval techniques for the analysis of dermatological lesions using particle swarm optimization technique, Appl. Soft Comput. 30 (2015) 650–662.

CHAPTER 2

Preprocessing and segmentation of skin lesion images

2.1 Introduction

Computer aided skin disease identification consists of the techniques implementing advanced mathematical tools to extract significant diagnostic features from the skin lesion images. For in-depth visualization and identification of the skin lesions, dermoscopic images are acquired from the affected areas. Prior to the feature extraction, proper segmentation of region of interest is an important and challenging job for the development of an indigenous system. In this context, removal of noise and artifacts from the lesion area are essential for further feature extraction and analysis of the disease.

Depending on various anatomical sites and image acquisition parameters, dermoscopic images contain different types of noise and artifacts. Improper lighting conditions, unnecessary movement of the patient, distance of the lens from the affected area, etc. sometimes introduce noise in the acquired dermoscopic images. Besides these sources of noise, significant portion of human skin covered with thick or thin hairs introduce artifacts in the lesion area. Presence of such hair like structures in the lesion area occludes some texture and boundary information from the lesion. Hair like objects lead to inappropriate segmentation of the region of interest. Introduction of an efficient noise removal and hair artifacts removal algorithm ensures proper segmentation and further feature extraction of the disease. Contemporary literature suggests various hair-removal techniques for the artifacts and noise removal step [1–5]. Lee et al. have introduced the DullRazor hair-removal algorithm, implementing bilinear interpolation technique [6]. Morphological operations and thresholding technique implemented in CIE L*u*v color space to detect and remove hair artifacts have been reported by Fleming et al. [7]. The research group led by Xie has proposed hair line detection algorithm followed by nonlinear partial differential equation based diffusion method for image inpainting and removal of hairs [8]. Abbas et al. have used derivative of Gaussian filter for the detection of hair like structures, morphological techniques for image refinement and fast marching image inpainting algorithm for the preprocessing of dermoscopic images [9].

Lesion segmentation is the primary requirement for computer aided diagnostic system. Proper segmentation of the affected area helps to determine the structural

Recent Trends in Computer-aided Diagnostic Systems for Skin Diseases
DOI: https://doi.org/10.1016/B978-0-323-91211-2.00002-0

property of the lesion. The appearance of skin lesions in dermoscopic images vary considerably according to the skin condition. Due to complex and irregular nature of the skin lesions, it is very difficult and challenging task to segregate the pigmented area from the normal skin region. The proper segmentation and border detection of the lesion is important to identify the structural irregularity of the lesion and further spreading of the disease. In literature, different thresholding based, region based, active contour based techniques are proposed for the segmentation of pigmented lesion area [10–13]. Glaister et al. have introduced texture based skin lesion segmentation algorithm calculating texture distinctiveness metric [14]. An unsupervised Delaunay Triangulation technique is used to segment the lesion region for the detection of melanoma in [15]. Flores et al. have introduced an adaptive unsupervised dictionary learning technique for the segmentation of melanocytic skin lesions [16]. Ma and Tavares have proposed a deformable model based effective and flexible segmentation approach for the segregation of skin lesion from dermoscopic images [17]. Guarracino and Maddalena have proposed SDI +, an unsupervised algorithm for extracting primary information on possible confounding factors for accurate segmentation of the lesion [18]. Li et al. have elaborated enhanced convolutional–deconvolutional networks with smaller convolutional kernels, incorporating color information from multiple color spaces, for segmentation of skin lesions [19].

In this chapter, mathematical morphology based techniques have been explored for artifacts removal and segmentation of the lesion from the dermoscopic images. Simple and efficient algorithms are developed for the extraction of hair like structures from the skin lesion area. Extraction and elimination of artifacts from the dermoscopic images assure proper segmentation of the lesion area. Segmentation algorithm implementing morphological operations is developed to segment wide variety of skin lesion types. In the subsequent sections, dermoscopic image preprocessing and lesion segmentation methodologies are explained extensively.

2.2 Mathematical morphology: a cursory view

Mathematical morphology is defined as a mathematical theory for the analysis of spatial structures. It aims at analyzing the shape and form of the objects based on set theory, lattice algebra and integral geometry. It is a powerful image analysis tool for investigating the interrelation between an image and a certainly chosen structuring element (SE) using some basic morphological operations. The morphological approach implementing basic morphological operator is useful for extracting relevant structures from the images. The shape of the SE is selected based on prior knowledge of relevant structures present in the image. Some of the basic morphological operations are erosion, dilation, opening, and closing. These basic operations are discussed in the subsequent subsections.

2.2.1 Structuring element

The fundamental element of mathematical morphology is SE. A SE is usually a small set relative to the image under study, with various sizes and shapes. The 2D SEs for 2D image processing are often referred to as flat SE. For investigating the morphology of n-dimensional image objects, $n+1$-dimensional SEs, called volumic, nonflat, or grayscale SEs are used. For a flat SE, the size of the matrix determines the size of the SE. Accordingly, the pattern of one-zero combination represents the shape of the SE. A common practice is to have odd dimensions of the structuring matrix and the origin is defined as the center of the matrix. The origin allows the positioning of the SE at a location on the image under consideration, coinciding the origin with that pixel of the image. The shape and size of the SE must be adapted to the structural properties of the image objects. For instance, linear SEs are used for the extraction of linear objects [20]. Some of the SEs are shown in Fig. 2.1.

2.2.1.1 Matlab functions to design different structuring element

A *strel* object or function is used to generate flat or binary morphological SE in Matlab. This function creates a 2D binary valued neighborhood of various shapes as described below. The flat or binary SEs can be used to perform various morphological operations with binary and grayscale images.

Syntax:

$$SE = strel(\text{“}shape,\text{”}size)$$

Strel creates SEs of various shapes like *“diamond,” “disk,” “square,” “line,” “rectangle,”* and so on. To generate SEs of different shapes, the *size* of the corresponding object must be specified by providing its *radius* or *length*.

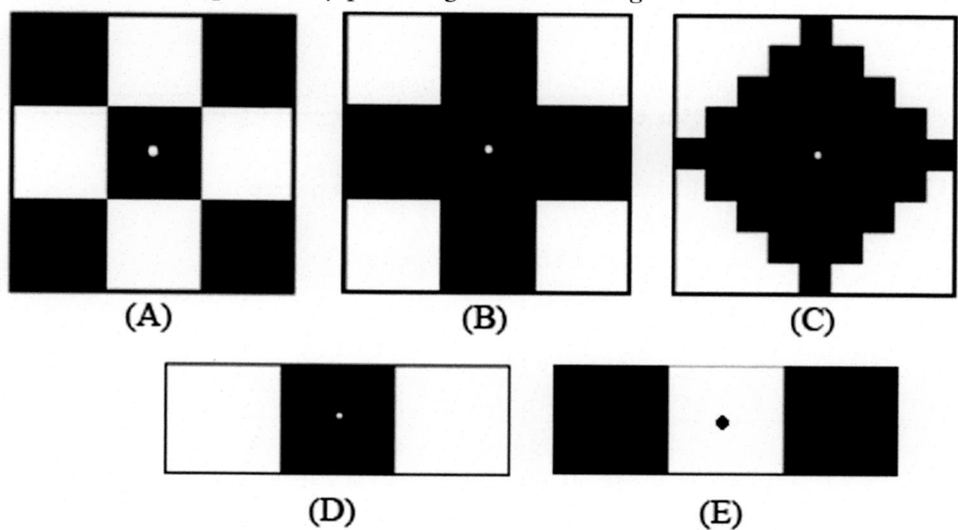

Figure 2.1 Examples of different binary structuring elements (SEs): (A) and (B) square, (C) diamond, (D) and (E) linear SEs.

SE = strel("diamond," r)—provides a *diamond*-shaped SE, where ***r*** corresponds to the distance from the origin of the SE to points of the diamond.
SE = strel("line," l, orient)—forms a symmetric linear SE of *length **l*** with an *angle* of *orient.*
SE = strel("rectangle," [m, n])—generates a rectangular SE of *size* $m \times n$. For symmetric SE (having an origin) the size should be chosen as odd number.
SE = strel("square," s)—creates a square SE of *length **s***. Similar to the rectangular SE, to obtain a symmetric square SE the *length* of the object should be selected as odd number. The square SE can be generated by using the same syntax as rectangular SE (selecting $m = n$).
Example:
Examples of some of the common SEs have been shown below.

- Create a diamond-shaped SE
  SE = strel("diamond," 15)
  SE = strel is a diamond-shaped SE with properties:
  Neighborhood: [31 × 31 *logical*]
  Dimensionality: 2
  imshow(SE.Neighborhood)—display the diamond-shaped SE

- Create a 11 × 11 square SE
  SE1 = strel("square," 11*)*
  SE1 = strel is a square-shaped SE with properties:
  Neighborhood: [11 × 11 *logical*]
  Dimensionality: 2
- Create a linear SE of length of 11 and an angle of 30 degree
  SE2 = strel("line," 11, 30)
  SE2 = strel is a line-shaped SE with properties:
  Neighborhood: [5 × 9 logical]
  Dimensionality: 2
  SE2.Neighborhood—display the SE

```
0 0 0 0 0 0 0 1 1
0 0 0 0 0 1 1 0 0
0 0 0 0 1 0 0 0 0
0 0 1 1 0 0 0 0 0
1 1 0 0 0 0 0 0 0
```

2.2.2 Erosion and dilation

The fundamental operation of mathematical morphology is erosion. The erosion of a set $\boldsymbol{X}$ by a SE $\boldsymbol{S}$ is denoted by $\varepsilon_s(X)$ and can be expressed as [20,21]:

$$\varepsilon_S(X) = \{x | S_x \subseteq X\} \tag{2.1}$$

The definition can be further extended to binary and grayscale images. The erosion of an image I by a SE $\boldsymbol{S}$ is defined as the minimum of the translations of $\boldsymbol{I}$ by the vectors $-\boldsymbol{s}$ of $\boldsymbol{S}$:

$$\varepsilon_S(I) = \underset{s \in S}{\vee} f_{-s} \tag{2.2}$$

Hence, the eroded value of a particular pixel i coinciding with the origin of the SE is the minimum value of the image region encountered in the window defined by the selected SE [20]:

$$[\varepsilon_S(I)](i) = \min_{s \in S} I(i + s) \tag{2.3}$$

If the origin comes within the SE, the erosion shrinks the input image which is a subset of the original image. Erosion eliminates the smallest connected components that are not encountered by the SE.

The dual operator of erosion is dilation. The dilation of a set $\boldsymbol{X}$ by a SE $\boldsymbol{S}$, denoted by $\delta_S(X)$, can be defined as [20,21]:

$$\delta_S(X) = \left\{x |, S_x \bigcap X \neq \varnothing\right\} \tag{2.4}$$

The above equation can be rewritten in terms of a union of a set as following;

$$\delta_S(X) = \bigcup_{s \in S} X_{-s} \tag{2.5}$$

Extending the definition to binary and grayscale image, the dilation of an image I by a SE S is defined as the maximum of the translation of I by vectors—s of S:

$$\delta_S(I) = \underset{s \in S}{\vee} I_{-s} \tag{2.6}$$

In other words, when the center of the SE coincides with a certain pixel (i) of the image (I), the dilated value of that pixel is the maximum value of the pixels appearing in the window of the SE [20,21]:

$$[\delta_S(I)](i) = \max_{s \in S} I(i + s) \tag{2.7}$$

2.2.2.1 Morphological erosion and dilation using Matlab

I = *imread("File name")*; %% Read image *I* if the source file is in the same MATLAB directory/path%%

or, *I* = *imread("Path\File name")*; %% Read image I if the source in other file/directory %%

imshow(I)%% display the image %%

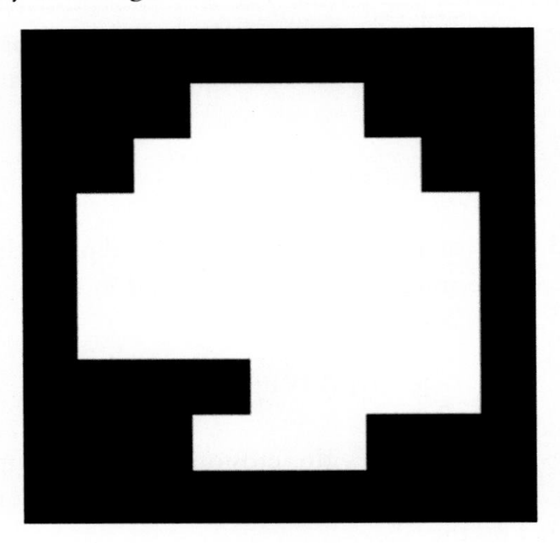

SE = *[0 1 0; 1 1 1; 0 1 0]*; %% Create the *SE*. SE can be generated using standard *strel* command or can be designed according to own choice.

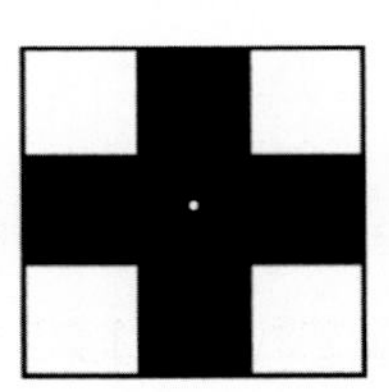

I_e = *imerode(I, SE)*; %% Morphological erosion operation on the image I with the *SE*.

figure, imshow(I_e) %% display the eroded image I_e.

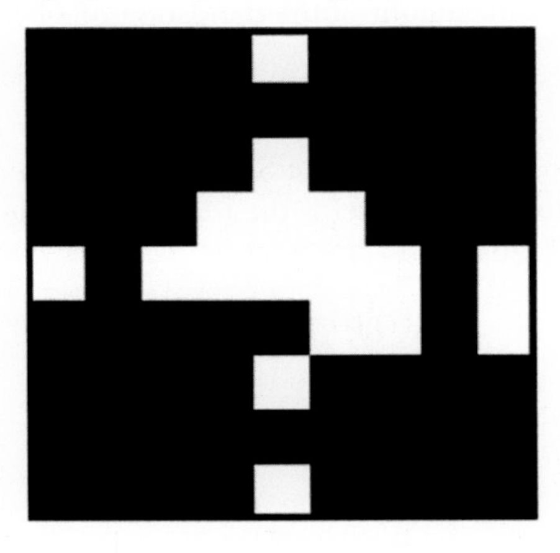

I_d = *imdilate(I, SE)*; %% *Morphological dilation* operation on the image *I* with the *SE*.

figure, imshow(I_e) %% display the eroded image I_e.

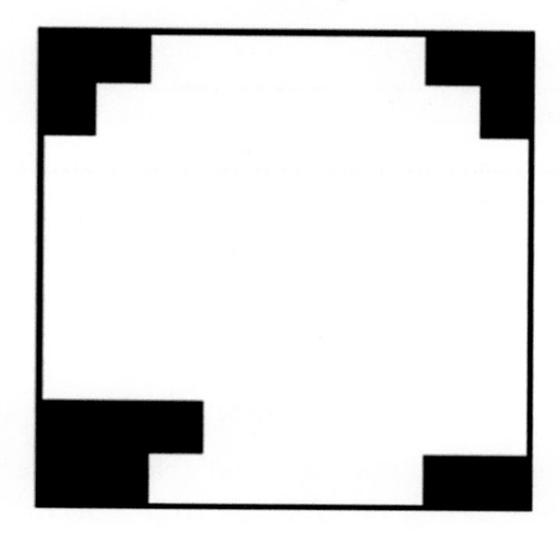

Examples of erosion and dilation are shown in above figures. The dilation results in an expansion of the image. Since dilation implicate a fitting into the complement of an image, it represents a filtering on the outer boundary region, whereas erosion represents a filtering on the inner boundary region. Dilation fills the small holes and regions of the image, comparable to the size of the SE, whereas erosion eliminates the smaller subregions. It is however noteworthy that owing to the nonlinear character of the operations, erosion is not the inverse of dilation. The inverse of either operation does not exist. Considering a symmetrical SE, the erosion and dilation holds the following properties [20,21].

- Erosion and dilation are dual operations.
- Both erosion and dilation are increasing operations.
- If the origin completely fills the SE, the dilation is extensive operation and erosion is antiextensive.
- 2D erosion and dilation can be carried out using one-dimensional erosions and dilations.

2.2.3 Morphological opening and closing

The composition of erosion and dilation operation introduces interesting properties of opening and closing. The erosion operation shrinks the image and removes the structures that are not encompassed by the SE. The eliminated structures cannot be recovered at all. To partially recover the structures lost by the erosion, morphological opening operation is introduced. Morphological closing is the dual operation of morphological opening. These operations are the basis of morphological filtering approach. The morphological opening ($\gamma_S(I)$), of an image (***I***) by a SE

S is defined as the erosion of (***I***) by ***S*** followed by the dilation with the reflected SE Š [21].

$$\gamma_S(I) = \delta_{\check{S}}[\varepsilon_S(I)] \tag{2.8}$$

Therefore, the idea behind the morphological opening is to recover the original image as much as possible by dilating the eroded image.

Similarly, morphological closing, denoted as $\varphi_S(I)$, can be defined as the dilation of an image (***I***) with a SE ***S*** followed by the erosion with the reflected SE Š [21].

$$\varphi_S(I) = \varepsilon_{\check{S}}[\delta_S(I)] \tag{2.9}$$

Examples of morphological opening and closing are shown in Fig. 2.3. Considering morphological closing of an image with a SE of a certain size, expanded boundaries of the region of interest after dilation can be partially incomplete by erosion followed by the dilation. In other words, dilation fills the small holes and thin tube like structures along the interior or at the boundary regions which are not reconstructed by the erosion. In this sense, closing can be explored to removes holes and thin cavities and opening to open up holes and remove small objects compared to the size of the SE [20,21].

Considering a symmetrical SE, the properties of opening and closing are listed below [20,21]:
- Opening and closing both are dual operations.
- Both Opening and closing operations are the combinations of erosion and dilation. So, similar to erosion and dilation, these are also increasing operations.
- Opening is antiextensive operation, whereas closing is extensive operation.
- Both opening and closing are idempotent operations, that is subsequent application of the operation on an image results the same.

$$\gamma\gamma(I) = \gamma(I) \tag{2.10}$$

$$\varphi\varphi(I) = \varphi(I) \tag{2.11}$$

Morphological opening and closing using Matlab

I = *imread("File name")*; %% Read image *I* if the source file is in the same MATLAB directory/path%%

or, *I* = *imread* ("*Path\File* name"); %% Read image I if the source in other file/directory %%

imshow (*I*)%% display the image %%

SE = [0 1 0; 1 1 1; 0 1 0]; %% Create the *SE*. SE can be generated using standard *strel* command or can be designed according to own choice.

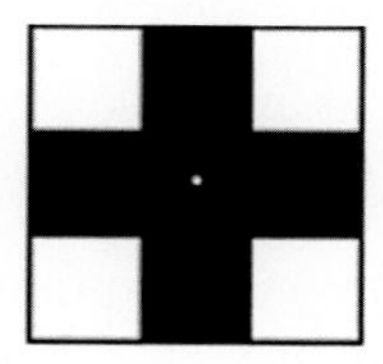

I_o = *imopen(I, SE)*; %% *Morphological opening* operation on the image *I* with the *SE*.
figure, imshow(I_o) %% display the eroded image I_o

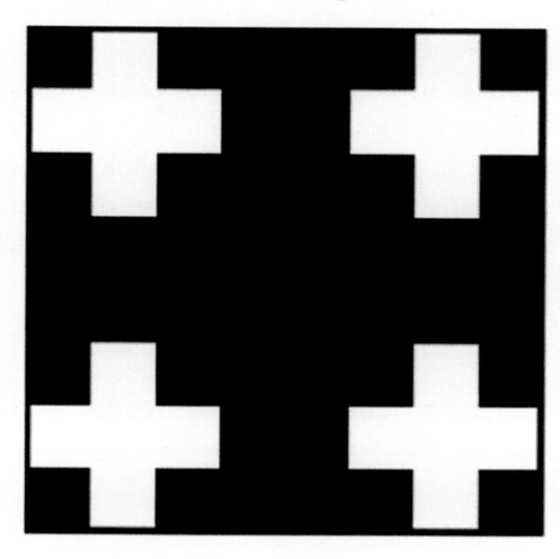

I_d = *imdilate(I, SE)*; %% *Morphological dilation* operation on the image *I* with the *SE*.

figure, imshow(I_e) %% display the eroded image I_e.

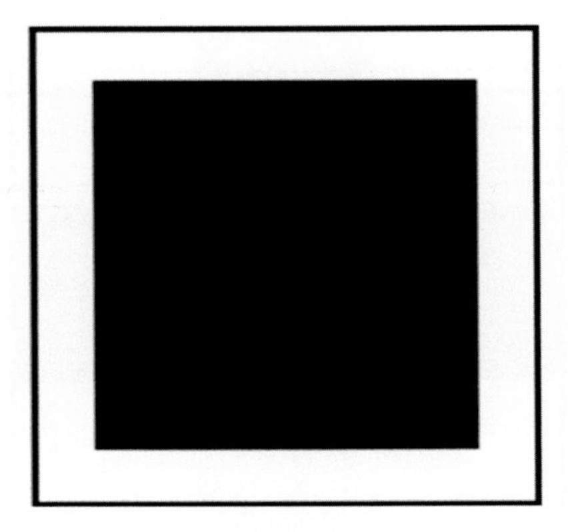

2.2.4 Residues

Residues are used for the formation of different morphological filters, evaluating the difference between two or more common morphological operations. Depending on various types of operations, some of the distinguishing residues are as the following [20]:

- morphological gradient,
- morphological Laplacian,
- top-hat filter,
- thinning and thickening using hit–miss transform.

Among the various types of residues, in this book, morphological gradient and top-hat operations are extensively used for the development of morphological filters.

2.2.4.1 Morphological gradient

In general, gradient of an image is its first derivative. For an n-dimensional image ($n > 1$), the gradient operation results an *n-vector* at each point. In morphological gradient operation the magnitude of the gradient is considered. The morphological gradient can be used to determine the local variations or variations along the edges of an image [20].

Considering erosion and dilation of an image (***I***), the morphological gradient *G(I)* can be defined as:

$$G(I) = \delta_S(I) - \varepsilon_S(I) \tag{2.12}$$

where ***S*** is the SE. This equation can be further extended as the *internal* and *external* gradient of the image (*I*) as the following [20,21]:

$$G^{-}(I) = I - \varepsilon(I) \tag{2.13}$$

$$G^{+}(I) = \delta(I) - I \tag{2.14}$$

where, the internal ($G^-(I)$) and external ($G^+(I)$) gradients adheres to the inside and outside of the objects, respectively.

In contrast to the internal and external gradient, the morphological Laplacian $\Delta(I)$ is defined as the residue of the external and internal gradient:

$$\Delta(I) = G^+(I) - G^-(I) \tag{2.15}$$

2.2.4.2 Top-hat filter

The selection of a morphological filter depends on the structural properties, shape, size, and orientation of the objects akin to filter. Depending on the size of the SE, opening or closing remove the small structures compared to the selected SE. These structures can be retrieved through the arithmetic difference between the image and its opening or between the closing and the image. These arithmetic differences are the basis for the development of morphological top hats [20,21].

The *white top-hat* (F_{WTF}), also called *top-hats by opening* of an image $\boldsymbol{I}$ is the difference between the original image $\boldsymbol{I}$ and the opening of the same [20,21]:

$$F_{WTH}(I) = I - \gamma(I) \tag{2.16}$$

$$F_{BTH}(I) = \varphi(I) - I \tag{2.17}$$

Since the opening is an antiextensive image transformation, the grayscale values of the *white top-hat* are always greater or equal to zero.

The counterpart of white top-hat filter is the *black top-hat* filter (F_{BTH}), which can be defined as the difference between the closing of the original image ($\boldsymbol{I}$) and the original image [20,21]:

$$F_{BTH}(I) = \varphi(I) - I \tag{2.18}$$

Black top-hat filter is also known as *top-hat by closing*. Black and white top hats are complementary operations.

2.2.4.3 Binary and gray-level hit-or-miss transform

The hit-or-miss transform (HMT) [22,23] is a classical morphological operation that uses two composite SE. To study the relationship between foreground and background of an image it is important to probe them at the same time. The HMT efficiently utilizes this concept.

If the two composite SEs are considered as S_f and S_b for foreground and background, respectively, and share the same origin, then $S_f \cap S_b = \emptyset$. $S = (S_f, S_b)$ denotes the composite SE.

The HMT of a binary image I by the composite SE S is the set of points i such that when the origin of S coincides with that of i S_f matches I while S_b matches I^c (the complement of I):

$$I \circledast S = (I \ominus S_f) \cap (I^c \ominus S_b) \quad (2.19)$$

Here, the operator $\circledast$ denotes the binary HMT and $\ominus$ denotes the binary erosion operation. The binary erosion of an image I with a SE S_f is defined as follows [24]:

$$I \ominus S_f = \{z | (S_f)_z \subseteq I\} \quad (2.20)$$

The equation indicates that the erosion of X with S_f is the set of all points z such that S_f, translated by z, is contained in I.

In other words, it can be said that the HMT is performed by translating the origin or center of the SE to all points in the image. The pixels of SE are compared with the image pixels that are covered by the SE. If the foreground and background pixels in the SE exactly match the pixels in the image covered by the SE, then the pixel underneath the origin or center of the SE is set to the foreground color otherwise to the background color.

As the binary HMT is not an increasing operation, it is somewhat difficult to generalize the HMT for gray-level image analysis. Many authors have introduced different approaches for gray-level hit-or-miss transform (GHMT) in the literatures [20,23].

Using the gray-level dilation and erosion, the GHMT can be defined as follows [23]:

$$GHMT(i) = max\left\{ \left(I \ominus S_f\right)(i) - (I \oplus S_b{}^c)(i), 0 \right\} \quad (2.21)$$

Here the operator $\ominus$ performs the grayscale erosion operation of the image X with the SE S_f. At any location (p, q), it is defined as the minimum value of the image in the region coincident with S_f when the origin of S_f is at (p, q) [24]:

$$\left[I \ominus S_f\right](p, q) = \min_{(m, n \in S_f)} \left\{ I(p + m, q + n) \right\} \quad (2.22)$$

Similarly, the dilation of the image I with the SE S_b^c denoted by the operator $\oplus$, at any location (p, q) is defined as the maximum value of the image obtained with the region contain by S_b^c when the origin of S_b^c is at (p, q) [24]:

$$[I \oplus S_b{}^c](p, q) = \max_{(m, n \in S_b{}^c)} \left\{ I(p - m, q - n) \right\} \quad (2.23)$$

2.2.4.3.1 HMT with multistructuring elements

The HMT extracts those objects that have shape and size similar to those of the SE. So, a properly chosen SE pair should be able to detect the linear objects of similar shape and size. To extract the objects of varying width, a multi-SE with varying width in different orientation can be selected for the GHMT. According to [25] most of the linear structures can be detected by lines in four directions as 0°, 45°, 90°, and 135°. Considering this, the orientation of the SE is to be varied from 0° to 165° with an increment of 15°. That is, within 0° to 165°, twelve directions is considered and the width of the SEs can also been varied in each orientation.

For the HMT, the foreground and the background SEs can be selected as follows:

$$S_f = \overbrace{\begin{bmatrix} 0 & 0 & 0 \\ 1 & 1 & 1 \\ 0 & 0 & 0 \end{bmatrix}}^{L_f} \qquad S_b = \overbrace{\begin{bmatrix} 1 & 1 & 1 \\ 0 & 0 & 0 \\ \vdots & \vdots & \vdots \\ 0 & 0 & 0 \\ 1 & 1 & 1 \end{bmatrix}}^{L_f} \Big\} W_b \tag{2.24}$$

Here, the length of the foreground SE (S_f) and the background SE (S_b) has been denoted by L_f and L_b, respectively. The width of S_b, that is, the rows of zeros, has been denoted as W_b. In these expressions, one represents the region to be covered and zero represents the regions that are uncovered. Here, each of the SE is rotated in steps of 15° angle spanning 0° to 165°. The rotation of the SE can be performed using affine transformation, that is, by using a transformation matrix $\mathbf{T} = \begin{bmatrix} \cos\theta & -\sin\theta & 0 \\ \sin\theta & \cos\theta & 0 \\ 0 & 0 & 1 \end{bmatrix}$. It rotates the SE counterclockwise by an angle θ.

Using the SEs (S_{f0}, S_{b0}), the HMT detects the linear regions corresponding to 0° angle as follows.

$$GHMT(S_{f0}, S_{b0})(i) = \max\{(I \ominus S_{f0})(i) - (I \oplus S_{b0}{}^{c})(i), 0\} \tag{2.25}$$

Similarly, the HMTs for 15°, 45°, 90°, and 165° rotation of SEs are given in the following:

$$GHMT(S_{f15}, S_{b15})(i) = \max\{(I \ominus S_{f15})(i) - (I \oplus S_{b15}{}^{c})(i), 0\} \tag{2.26}$$

$$GHMT(S_{f45}, S_{b45})(i) = \max\{(I \ominus S_{f45})(i) - (I \oplus S_{b45}{}^{c})(i), 0\} \tag{2.27}$$

$$GHMT(S_{f90}, S_{b90})(i) = \max\{(I \ominus S_{f90})(i) - (I \oplus S_{b90}{}^{c})(i), 0\} \tag{2.28}$$

$$GHMT(S_{f165}, S_{b165})(i) = \max\{(I \ominus S_{f165})(i) - (I \oplus S_{b165}{}^{c})(i), 0\} \tag{2.29}$$

For each degree of rotation, the HMT is to be performed by considering the maximum operation and finally all the extracted linear features can be combined to extract every possible linear region present in an image in all possible directions:

$$I_l = GHMT(S_{f0}, S_{b0})(i) + GHMT(S_{f15}, S_{b15})(i) + \cdots + GHMT(S_{f165}, S_{b165})(i) \tag{2.30}$$

For the extraction of all the possible linear objects, it is expected to perform the hit-or-miss with varying widths of the pair of SEs. In this section, the width of the background SE (S_b), W_b is varied in each orientation, keeping the width of the foreground SE (S_f) fixed.

If n number of scales are used, then the width of the background SE W_b is varied as $W_b = 2 \times i + 1$; for $1 \leq i \leq n$. Then the detected linear regions corresponding to direction 0° at scale i, can be expressed as follows:

$$GHMT(S_{f0}, S_{b(i,0)})(i) = \max\left\{(I \ominus S_{f0})(i) - (I \oplus S_{b(i,0)}{}^{c})(i), 0\right\} \tag{2.31}$$

Similarly, the HMT is performed for each orientation with different scales by the pair of SEs as $(S_{f15}, S_{b(i,15)})$, $(S_{f30}, S_{b(i,30)})$, $(S_{f45}, S_{b(i,45)})$ and so on. Finally, the detected linear features at scale i are expressed in the following manner:

$$I_l^i = GHMT(S_{f0}, S_{b(i,0)})(i) + GHMT(S_{f15}, S_{b(i,15)})(i) + \cdots + GHMT(S_{f180}, S_{b(i,180)})(i) \tag{2.32}$$

The reported HMT can be used for the extraction of both bright and dark linear regions of an image. The image can be subjected to HMT with multiscale multi-SE, to identify the linear structures present in the image. Therefore, to ensure the extraction of all the brighter regions, the maximum operation is performed, considering all the resultant images from the HMT to segment the entire possible linear structures present in the original image. According to Eq. (2.14), the final segmented image I_l has been obtained by considering the maximum operation on all the resultant images obtained from HMT, denoted as $I_l^1, I_l^2, I_l^3, \cdots, I_l^n$:

$$I_l = \max\left\{I_l^1, I_l^2, I_l^3, \cdots, I_l^n\right\} \tag{2.33}$$

2.3 Application of noise and hair-removal techniques

The pre-processing techniques are employed for image noise and artifacts removal and to support better generalization ability. The different stages of preprocessing comprise of image color standardization, normalization, resizing, noise elimination and hair artifacts removal. An efficient preprocessing algorithm ensures improved performances in the subsequent stages of lesion segmentation, feature extraction and classification of the diseases.

2.3.1 Color standardization

The dermoscopic images can be acquired using different dermoscopes with various lighting conditions. The color variation present in the images due to uneven illumination of light is corrected and normalized using the gray world color constancy algorithm. The aim of the color constancy is to adjust the color of the acquired image under unknown light source. This transformation makes uniform distribution of colors under that light source. In this methodology, to accomplish the color transformation, the source of the light is estimated in the RGB color plane, $[e_R e_G e_B]$ and consequently estimate the illuminant from the transformed image. Considering the gray world color constancy algorithm, the each component of illuminant for a color image is estimated using the following expression [26,27]:

$$\frac{\int I_c(x)dx}{\int dx} = ke_c \tag{2.34}$$

where, I_C corresponds to the cth component of the image I, $x = (x, y)$ is the spatial location of a pixel and k is the normalization constant that ensures $e = [e_R e_G e_B]^T$ encompassing unit length according to the Euclidean norm.

2.3.2 Normalization

In the image acquisition stage, the dermoscopic images are acquired from different dermoscopes having different settings. To increase the dynamic range and to bring the range of intensity values of the images to normal distribution, image normalization technique is employed after the color standardization step. Here, the computed mean RGB values over the entire dataset, considered in this study is subtracted from each of the dermoscopic images [27,28].

2.3.3 Removal of noise and hair artifacts

After color standardization and normalization step, median filter is employed to reduce random noise present in the dermoscopic images. Median filter, a nonlinear filter is used to obtain a filtered image accomplishing a window sliding throughout the image. The filtered image is obtained by replacing the center pixel value with the median value estimated from the neighboring pixels in the image region encountered by the window. Considering the comparatively uniform regions of the image, the corresponding gray value is estimated by the median filter. As an edge is crossed, one side or the other dominates the window, and the output switches sharply between the values. Thus, the edge is not blurred. To preserve the edge information of an image, median filter has been considered for the removal of noise from dermoscopic images.

Most of the dermoscopic images are contaminated with hair artifacts [29]. The proper removal of the hair like structures from the original grayscale images led to the accurate segmentation of the skin lesion area. The removal of thick hairs is a challenging task as it may produce some hair shadow in the original image. The existence of the hair shadow introduces some unnecessary extension of the lesion area. As a result, the structure of the segmented object is changed. This section illustrates different morphological operations for the detection and elimination of the hair artifacts from the skin lesion area. The proper selection of shape and size of the SE ensures extraction of appropriate objects from the region of interest. The feature based object detection using morphological operations have been considered for accurate detection and further elimination of hair like structures. Morphological bottom-hat filtering [30], identifies those structures which are darker than their surroundings. To extract the hair like structures the morphological bottom-hat filter is employed with circular SE having radius of six pixels. It has been experimentally observed that the selection of SE of six pixels have extracted the thick and thin hair objects and removed without any hair shadow. The extraction and removal of hair like structures from the grayscale dermoscopic images have been shown in Fig. 2.2. From Fig. 2.2 it has been observed that

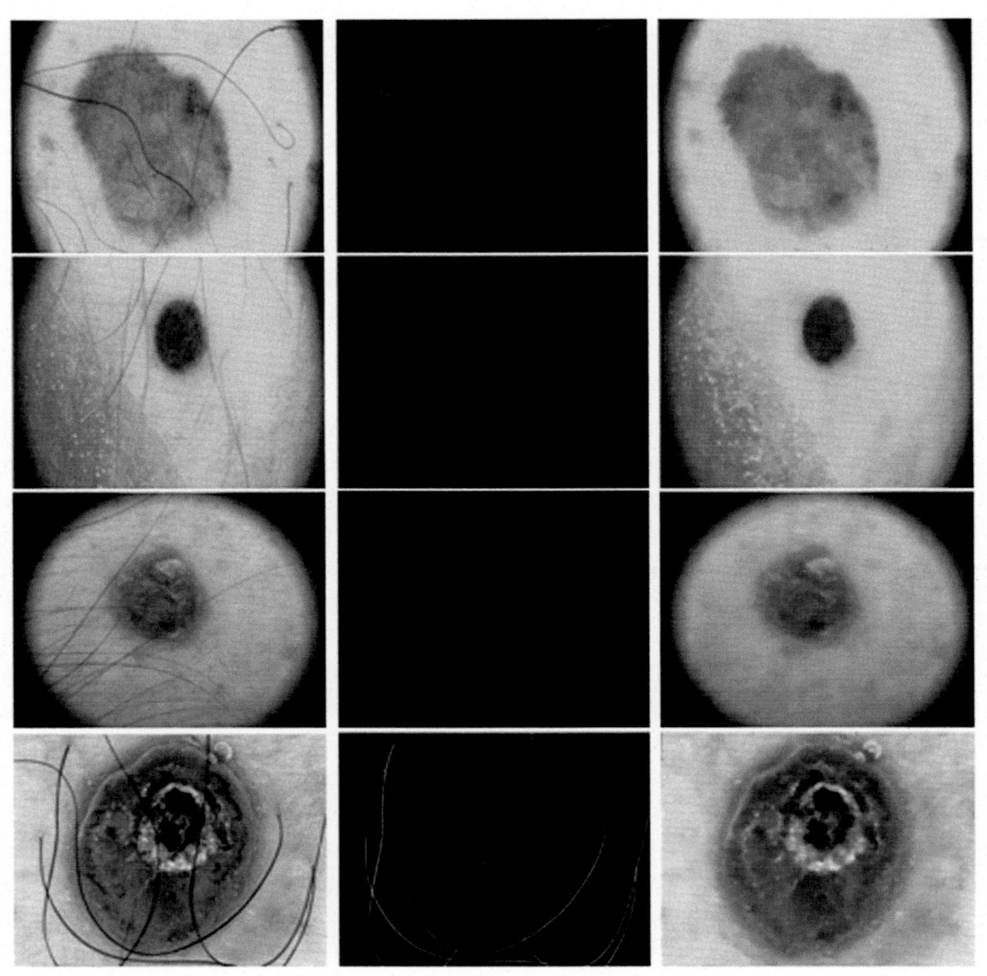

Figure 2.2 Performance of the hair-removal technique. Left column: original image with hair artifacts; middle column: extracted hairs, right column: hair removed images.

the circular SE helps to extract all the hair like structures of different width and orientation from the grayscale image. In the original grayscale image, the hair like structures manifesting as masks have been removed using inward interpolation method. It interpolates to extract the pixel values in the inside regions from those in the outer boundaries of the regions, by means of computing the discrete Laplacian over the regions and solving the Dirichlet boundary value problem, and efficiently eliminate the hair like objects from the original grayscale image.

2.4 Application of morphological techniques for skin lesion segmentation and border detection

The prescreening systems or an automatic diagnostic system for the diagnosis of skin abnormalities depend on the proper segmentation of the lesion area from the dermoscopic image. The proper segmentation of skin lesion area helps the dermatologists and experts to identify its morphological properties for further decision making. Segmentation of the affected area is essential to monitor the further spreading of the disease in a condition monitoring system. From the segmented image, the lesion border is detected to perceive the structural variations of the skin lesion. For an efficient computer aided diagnostic system, development of accurate lesion segmentation technique is very much challenging. The visual appearance of the skin lesion considerably varies according to the skin condition, texture and color. Some of the lesions have irregular and complex structure and in some cases there exist a smooth transition between the affected area and the normal skin. In this section, mathematical morphology has been explored for the proper segmentation and border detection of the skin lesion.

2.4.1 Materials

In the segmentation performance analysis study, primarily 6579 number of digital dermoscopic images, consisting of 2419 melanomas, 3261 dysplastic nevi and remaining 899 basal cell carcinoma (BCC) have been composed from the available datasets [31–34]. Dermoscopic images of these three disease classes have been shown in Fig. 2.3. The expert dermatologists have histopathologically confirmed the images considered for analysis. Among these three disease categories, melanoma lesions have very complex and irregular structure. For the BCC images, the color variation among the lesion area and normal skin is sometimes very difficult to differentiate. BCC lesions contain various morphologies consisting of dotted or linear vascular structures. The presence of this morphological properties makes it difficult to segment the BCC lesions from the dermoscopic images. Although the structures of the nevus are not much complex, diminutive structure with minimum intensity variation in the border region of lesions sometimes make segmentation a challenging task. Considering these

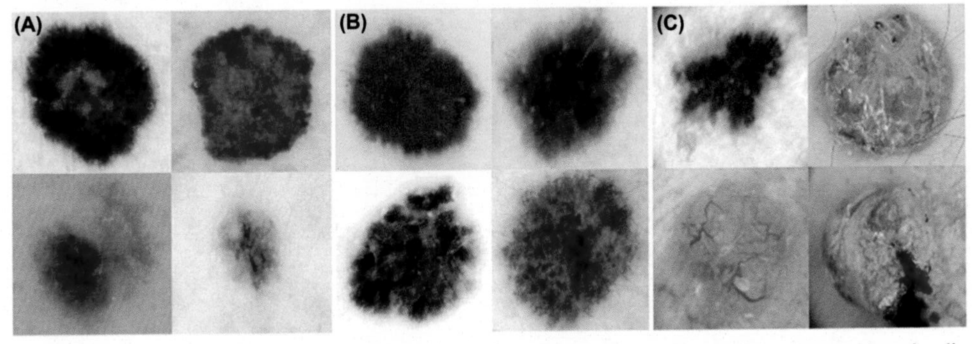

Figure 2.3 Examples of some dermoscopic images, (A) malignant melanoma, (B) benign nevus, (C) common basal cell carcinoma [34].

three types of skin diseases the segmentation problem is easy and not so easy depending on the type of skin lesions. The database also provides the ground truth (GT) segmented images for each of the corresponding original images. The GT images have been constructed by annotating the lesion area by the expert dermatologists. The GT images have been provided in the database as binary images of same size as the original images, where the white pixels have been considered as region of interest and black pixels as background.

2.4.2 Methodology

For the segmentation of melanoma, nevus, and BCC lesions from the dermoscopic images, morphological operations have been explored here. For the morphological operations, selection of SE is critical and depends on the structure of the object under consideration. As most of the skin lesions are circular in nature, circular kernel has been chosen as the SE for morphological operations. The radius of the circular kernel has been selected as four in terms of number of pixels to fit within most of the lesions. For the single pixel border detection, the kernel size has been determined as two-pixels in diameter, as larger kernel will widen the border region. The corresponding algorithm to generate circular SE is given in Algorithm 2.1.

On the grayscale dermoscopic image the morphological closing operation has been performed with circular SE. On subtracting the complement of the original grayscale image from the morphologically closed image, the region of interest that is the lesion area darker than its surrounding, has been obtained. A sharp contrast between the lesion area and its surrounding background has been obtained by this operation. To yield the segmented skin lesion area from the resultant image obtained after the morphological operations, the threshold value has been selected by minimizing the inter class variance of the image. After obtaining the histogram and the probability of each intensity level, the probabilities, means and variances of the two classes have been calculated for each of the threshold values from one to maximum intensity present in the image. The desired threshold corresponds to the maximum interclass variance of the resultant image.

To remove the dark corners in the original image, the segmented image has been masked with a circular binary mask. The skin lesions are not always in the center region of the original image and have irregular shape. To create this binary circular mask, the centroid of the segmented lesion has been considered as the center of the circle. The diameter of the circle has been varied according to the major axis length of the segmented lesion and has been made larger than that. After formation of this binary circular mask of varying position and diameter, it has been multiplied by the segmented image. The diameter of the circle has been varied until the ratio of the previous segmented lesion area and the masked lesion area is unity. As already stated, for melanoma identification, revealing the lesion border is of utmost important.

Algorithm 2.1:
Circular SE

Input: Radius r
1. SE = zeros($2 \times r + 1, 2 \times r + 1$)
2. *for* $i = 1$ to $2 \times r + 1$
3. *for* $j = 1$ to $2 \times r + 1$
4. Compute SE (i,j) = $\sqrt{(i-(r+1))^2 + (j-(r+1))^2}$
5. *if* $SE \leq r$
6. $SE(i, j) = 1;$
7. *else* $SE(i, j) = 0;$
8. *end if*
9. *end*
10. *end*

Output: Circular SE

Morphological gradient operation can be used for border detection, but it usually widens the lesion border. In this study, a SE circular in shape, having a diameter of two-pixels first erodes the segmented image. The eroded image is then subtracted from the original segmented image. This method has identified the pixel locations where the maximum change in intensity has occurred. The segmentation and border detection algorithm has been elaborated in Algorithm 2.2. Examples of segmentation and border detection of some dermoscopic images have been shown in Fig. 2.4.

Algorithm 2.2
Image segmentation and border detection

Input: Grayscale image I, $SE \leftarrow$ *Circular Kernel with* 8 *pixels diameter*
1: $I_c \leftarrow Morphological\_Closing(I, SE)$
2: $I_r \leftarrow Subtract(I_c, complement(I))$
3: *Compute histogram and probabilities of each intensity level*
4: *for all possible thresholds* $t = 1...max(I)$
Compute Probabilities, mean, variance
5: $Threshold(t_f) \leftarrow max(variance)$
6: $I_b \leftarrow Binary\_Thresholding(I_r, t_f)$
7: $I_m \leftarrow Circular_{BinaryMask}$, $diameter > majoraxislength(I_b)$
8: $I_{seg} \leftarrow masked(I_b, I_m)$
9: *Structuring Element* $SE \leftarrow$ *Circular Kernel with* 2 *pixels diameter*
10: $I_{er} \leftarrow morphological\_erosion(I_{seg}, SE)$
11: $I_{brdr} \leftarrow subtract(I_{seg}, I_{er})$;

Output: Segmented image I_{seg}, Border detected image I_{brdr}.

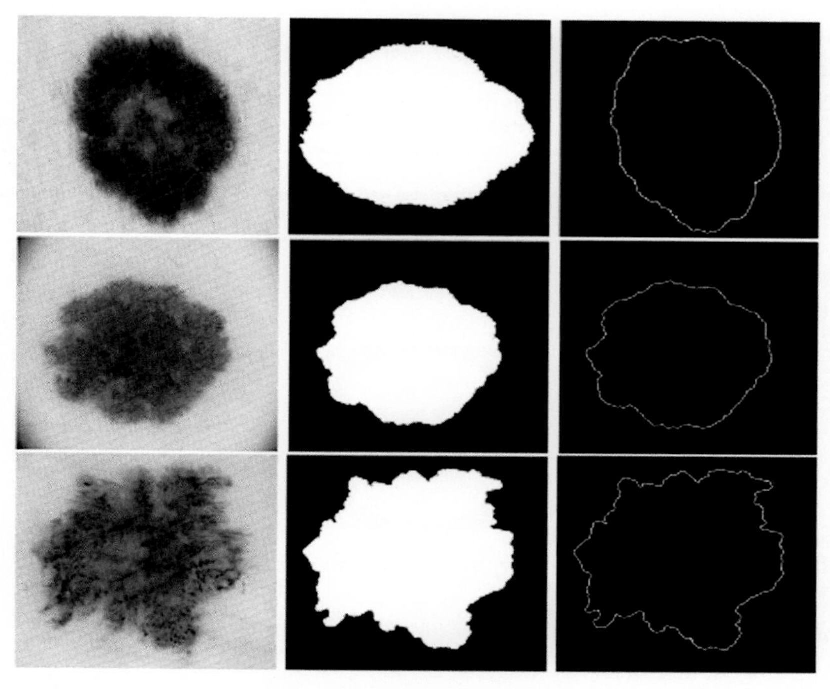

Figure 2.4 Segmentation and border detection of dermoscopic images. Left column: original images; middle column: segmented images, right column: border detected images.

2.4.2.1 Segmentation of skin lesion using Matlab

I = *imread("File name")*; %% Read image *I* if the source file is in the same MATLAB directory/path%%

or, *I* = *imread("Path\File* name"*)*; %% Read image I if the source in other file/ directory %%

I_g = *rgb2gray(I)*; %% Convert the RGB image into grayscale %%

imshow(I_g) %% display the image %%

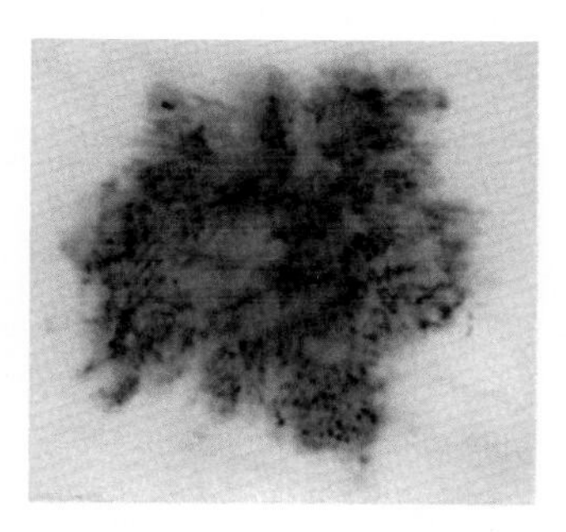

```
Create the circular SE:
  function [SE] = circularse(r)
    SE = zeros(2 × (r + 1), 2 × (r + 1))
    for i = 1: 2 × r + 1
     for j = 1: 2 × r + 1
       SE(i, j) = sqrt((i−(r + 1)).^2 + (j−(r + 1)).^2) ≤ r
     end
    end
  end
```

```
IC = imclose(Ig, SE); %% Morphological closing with circular SE
IG = IC—imcomplement(Ig)
figure, imshow(IG)
```

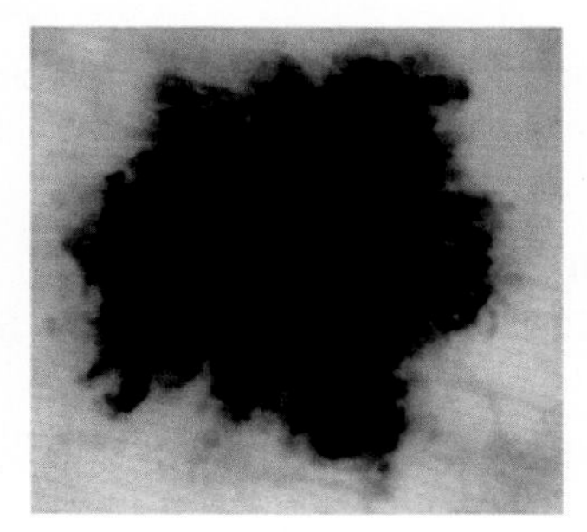

```
Compute threshold value T for image IG
Ib = im2bw(IG, T) %% Threshold image IG to obtain the Segmented image Ib %%
figure, imshow(Ib)
```

2.4.3 Results and discussion

The performance of the proposed algorithm in carrying out segmentation, has been evaluated by the quantitative estimation of the similarity between the segmented image and the corresponding GT image. The segmentation performance has been assessed by estimating the pixel level sensitivity (Sen), the specificity (Spec), the accuracy (ACU), along with the image similarity measuring indices, for example, Jaccard similarity index (JSI) and Dice similarity coefficient (DSC). The segmentation performance indices have been defined as the following:

The *sensitivity (Sen)* determines how the algorithm identifies the foreground pixels (*True Positive (TP)*) as foreground:

$$Sensitivity(Sen) = \frac{TP}{TP + FN}$$

The *specificity (Spec)* determines how the algorithm identifies the background pixels (*True Negative (TN)*) as background:

$$Specificity(Spec) = \frac{TN}{TN + FP}$$

The pixel label *accuracy (ACU)* for segmentation determines how the foreground and background pixels are identified as corresponding foreground and background:

$$Accuracy(ACU) = \frac{TP + TN}{TP + TN + FP + FN}$$

Here, *TP, TN, FP*, and *FN* denote the number of true positive, true negative, false positive, and false negative pixels, respectively.

The *JSI* of two sets is expressed as [35]:

$$JSI(I, G) = |I \cap G|/I \cup G$$

Where $|I|$ represents the cardinality of set $\boldsymbol{I}$. The *JSI* can be expressed in terms of *TP*, *FP*, and *FN* as:

$$JSI(I, G) = \frac{TP}{TP + FP + FN}$$

Similarly, the *DSC* of two sets $\boldsymbol{I}$ and $\mathbf{G}$ is expressed as [36]:

$$DSC = 2 \times |I \cap G|/(|I| + |G|)$$

The *DSC* can be expressed in terms of *TP*, *FP* and *FN* as the following:

$$DSC(I, G) = 2 \times \frac{TP}{TP + FP + FN}$$

Table 2.1 Skin lesion segmentation performance considering the melanoma, nevus, and BCC dermoscopic images of ISIC and PH2 datasets.

Database	Values	Segmentation performance indices				
		Sen	Spec	ACU	JSI	DSC
ISIC dataset	Min	0.6116	0.6228	0.7274	0.6157	0.6565
	Max	0.9996	0.9930	0.9963	0.9839	0.9919
	Avg	0.9172	0.9788	0.9521	0.8562	0.9142
PH2 dataset	Min	0.6121	0.6409	0.7144	0.6811	0.7594
	Max	1.00	1.00	0.9962	0.9597	0.9794
	Avg	0.9134	0.9510	0.9396	0.8060	0.8889

Sen, Sensitivity; *Spec*, specificity; *ACU*, accuracy; *JSI*, Jaccard similarity index; *DSC*, Dice similarity coefficient.

The segmentation performance indices have been tabulated in Table 2.1 for each of the ISIC and PH2 datasets individually [33,34], comprising of melanoma, nevus (both the datasets) and BCC (ISIC dataset) images. Table 2.1 has indicated the minimum, maximum and average values of all the performance indices. The segmentation performance indices have been evaluated considering the segmented lesion and the corresponding expert's annotated GT images.

The segmentation results in Fig. 2.5 and performance indices in Table 2.1 have revealed the effectiveness of the algorithm for skin lesion segmentation. The similarity between the segmented image and the corresponding GT is depicted with pink and green color, where the oversegmented and undersegmented regions correspond to green and pink color. Here, the mathematical morphology aided segmentation algorithm has achieved an acceptable outcome in skin lesion area detection. The reported algorithm has successfully segmented the complex structures of melanoma and BCC images along with comparatively easier lesion of nevus class. The similarity measuring indices of segmented image and its corresponding GT have justified the acceptable performance of the proposed algorithm. For the images of ISIC dataset, the average values of DSC and JSI similarity indices have reported considerably closer similarity values as 0.8562 and 0.9142, respectively. The segmentation algorithm has achieved pixel level sensitivity, specificity and accuracy of 0.9172, 0.9788, and 0.9521 for ISIC dataset. Similarly, values of 0.8060 for the DSC and 0.8889 for the JSI similarity indices have been obtained for the images of PH2 dataset. The noisy and poor quality dermoscopic images having soft difference between the lesion area and skin, have led to the improper segmentation of some of those dermoscopic images.

The performance of this segmentation algorithm has been compared with the recently published state-of-the-art segmentation techniques in Table 2.2. Dey et al. have reported the segmentation similarity indices JSI and DSC as 0.8192 and 0.8916, respectively, using localized active contour segmentation approach [35]. In [37], Rajanikanth et al. have segmented the dermoscopic images using Kapur's entropy and

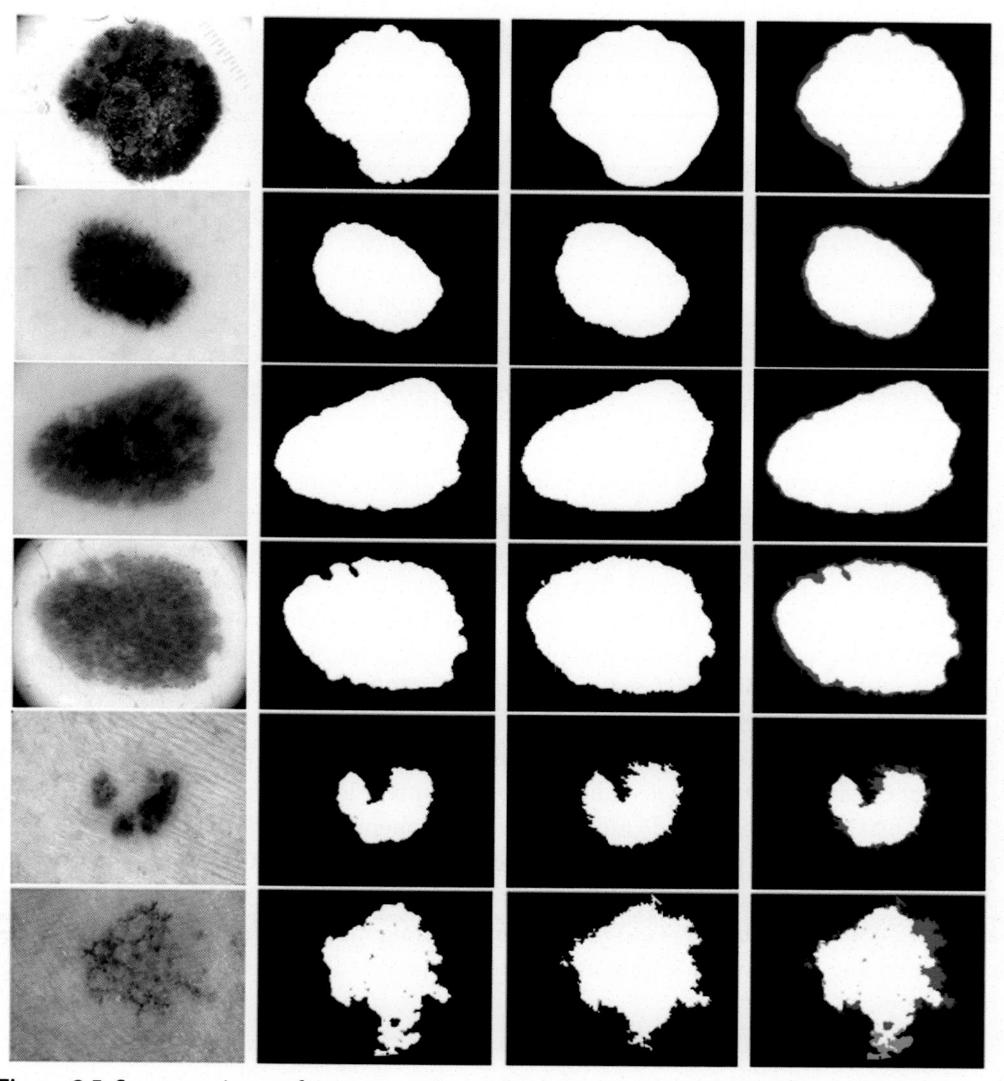

Figure 2.5 Segmentation performance analysis of the proposed methodology. First column: original images; second column: segmented images; third column: ground truth (GT) images; fourth column: similarity measure between the GT images and the resultant segmented images.

level set method with pixel level sensitivity of 0.9927, 0.9177 specificity, 0.9628 accuracy along with JSI and DSC of 0.8805 and 0.9138, respectively. M.A. Al-masni et al. have achieved the segmentation performance indices as 0.8540 sensitivity, 0.9669 specificity, 0.9403 of pixel level accuracy, 0.7711 JSI and 0.8708 DSC, employing deep full resolution convolution networks [38]. Garcia-Arroyo et al. have reported the

Table 2.2 Skin lesion segmentation performance comparison of proposed method with state-of-the-art methods.

Work	Segmentation performance indices				
	Sen	Spec	ACU	JSI	DSC
Dey et al. [35]	–	–	–	0.8192	0.8916
Rajanikanth et al. [37]	0.9927	0.9177	0.9628	0.8805	0.9138
M.A. Al-masni et al. [38]	0.8540	0.9669	0.9403	0.7711	0.8708
Garcia-Arroyo and Garcia-Zapirain [39]	0.8690	0.9230	0.9340	0.7910	0.8690
Bi et al. [40]	0.8020	0.9850	0.9340	0.7600	0.8440
Navarro et al. [41]	–	–	0.9550	0.7690	0.8540
Li and Shen [42]	–	–	0.9590	0.8700	0.931
Present methodology	0.9172	0.9788	0.9521	0.8562	0.9142

skin lesion segmentation performance indices of 0.8690 sensitivity, 0.9230 specificity, 0.9340 accuracy, 0.7910 JSI and 0.8690 DSC, using the algorithm based on fuzzy classification of pixels and subsequent histogram thresholding methodology [39]. Using deep residual network, Bi et al. have reported 0.8020 sensitivity, 0.9850 specificity, 0.9340 accuracy, 0.7600 JSI and 0.8440 DSC for skin lesion segmentation [40]. Navarro et al. have described super-pixel based skin lesion segmentation using local features and obtained pixel level accuracy of 0.9590, JSI of 0.7690 and a DSC of 0.8540 [41]. Li and Shen [42] have performed automatic delineation of skin lesion contours from dermoscopic images using dense deconvolutional network and obtained the segmentation performance indices as 0.9550 pixel level accuracy, 0.8700 JSI and 0.9310 DSC. The proposed morphological segmentation algorithm has achieved acceptable pixel level performance indices of 0.9172 sensitivity, 0.9788 specificity, and 0.9521 accuracy along with highly satisfactory similarity measure with the GT images as 0.8562 JSI and 0.9142 DSC. The segmentation algorithm, although simple has outperformed the skin lesion segmentation performances of others using deep learning techniques.

2.5 Conclusion

In chapter, mathematical morphology is extensively explored for preprocessing and segmentation of dermoscopic images. In the preprocessing stage, morphological filter is employed for the removal of hair artifacts. After the preprocessing stage, morphological gradient filter is designed with circular SE. The presented gradient operation detects the intensity variation among the lesion and normal skin area to segregate the lesion area from the dermoscopic images. Mathematical morphology based segmentation methodology has segmented the lesion area with 91.72% sensitivity, 97.88%

specificity and 95.21% accuracy. The performance of the technique is also evaluated by means of estimating the similarity measure of Jaccard index (JI) and DSC. The reported methodology has achieved acceptable indices of 85.62% JI and 91.42% DSC.

References

[1] D.H. Chung, G. Sapiro, Segmentation skin lesions with partialdifferential equation-based image processing algorithms, IEEE Trans. Med. Imaging 19 (2000) 763–767.

[2] C.A.Z. Barcelos, V.B. Pires, An automatic based non-linear diffusion equations scheme for skin lesion segmentation, Appl. Math. Comput. 215 (2009) 251–261.

[3] P.S. Saugeona, J. Guillodb, J.P. Thiran, Towards a computer-aided diagnosis system for pigmented skin lesions, Comput. Med. Imag. Grap. 27 (2003) 65–78.

[4] H. Zhou, M. Chen, R. Gass, J.M. Rehg, L. Ferris, J. Ho, L. Dragowski, Feature-preserving artifact removal from dermoscopy images, in: Proceedings of the SPIE Medical Imaging, pp. 1–9, 2008.

[5] P. Wighton, T.K. Lee, M.S. Atkinsa, Dermoscopic hair disocclusion using inpainting, in: Proceedings of the SPIE Medical Imaging, pp. 1–8, 2008.

[6] T.K. Lee, V. Ng, R. Gallagher, A. Coldman, D. McLean, A. Dullrazor, Software approach to hair removal from images, J. Comput. Biol. Med. 27 (1997) 533–543.

[7] M. Fleming, C. Steger, J. Zhang, J. Gao, A. Cognetta, I. Pollak, C. Dyer, Techniques for a structural analysis of dermatoscopic imagery, Comput. Med. Imag. Grap. 22 (1998) 375–389.

[8] F.Y. Xie, S.Y. Qin, Z.G. Jiang, R.S. Meng, PDE-based unsupervised repair of hair-occluded information in dermoscopy images of melanoma, Comput. Med. Imag. Grap. 33 (2009) 275–282.

[9] Q. Abbas, M.E. Celebi, I.F. García, Hair removal methods: a comparative study for dermoscopy images, Biomed. Signal Process. Control. 6 (2011) 395–404.

[10] R.B. Oliveira, M.E. Filho, Z. Ma, J.P. Papa, A.S. Pereira, J.M.R.S. Tavares, Computational methods for the image segmentation of pigmented skin lesions: a review, Comput. Methods Prog. Biomed. 131 (2016) 127–141.

[11] M.E. Celebi, H.A. Kingravi, B. Uddin, H. Iyatomi, Y.A. Aslandogan, W.V. Stoecker, R.H. Moss, A methodological approach to the classification of dermoscopy images, Comput. Med. Imaging Graph. 31 (2007) 362–373.

[12] J. Glaister, R. Amelard, A. Wong, D. Clausi, MSIM: multistage illumination modeling of dermatological photographs for illumination corrected skin lesion analysis, IEEE Trans. Biomed. Eng. 60 (2013) 1873–1883.

[13] B. Erkol, R.H. Moss, R. Joe Stanley, W.V. Stoecker, E. Hvatum, Automatic lesion boundary detection in dermoscopy images using gradient vector flow snakes, Skin Res. Technol. 11 (2005) 17–26.

[14] J. Glaister, A. Wong, D.A. Clausi, Segmentation of skin lesions from digital images using joint statistical texture distinctiveness, IEEE Trans. Biomed. Eng. 61 (4) (2014) 1220–1230.

[15] A. Pennisi, D.D. Bloisi, D. Nardi, A.R. Giampetruzzi, C. Mondino, A. Facchiano, Skin lesion image segmentation using Delaunay Triangulation for melanoma detection, Computerized Med. Img Graph. 53 (2016) 89–103.

[16] E. Flores, J. Scharcanski, Segmentation of melanocytic skin lesions using feature learning and dictionaries, Expert Syst. Appl. 56 (2016) 300–309.

[17] Z. Ma, J.M.R.S. Tavares, A novel approach to segment skin lesions in dermoscopic images based on a deformable model, IEEE J. Biomed. Health Inform. 20 (2) (2016) 615–623.

[18] M.R. Guarracino, L. Maddalena, SDI + : a novel algorithm for segmenting dermoscopic images, IEEE J. Biomed. Health Inform. 23 (2) (2018) 481–488.

[19] H. Li, X. He, F. Zhou, Z. Yu, D. Ni, S. Chen, T. Wang, B. Lei, Dense deconvolutional network for skin lesion segmentation, IEEE J. Biomed. Health Inform. 23 (2) (2019) 527–537.

[20] P. Soille, Morphological Image Analysis Principles and Applications, second ed., Springer, 2004.

[21] E.R. Dougherty, R.A. Lotufo, Hands-on morphological image processing, SPIE Publications (2003).

[22] J. Serra, Image Analysis and Mathematical Morphology, Academic Press, London, 1982.

[23] B. Naegel, N. Passat, C. Ronse, Gray-level hit-or-miss transforms—part I: unified theory, Patt. Recog. 40 (2007) 635–647.

[24] R.C. Gonzalez, R.E. Woods, Digital Image Processing, third ed., Pearson, 2014.

[25] X. Bai, T. Wang, F. Zhou, Linear feature detection based on the multi-scale, multi-structuring element, gray-level hit-or-miss transform, Comput. Elect. Eng 46 (2015) 487–499.

[26] C. Barata, M.E. Celebi, J.S. Marques, Improving dermoscopy image classification using color constancy, IEEE J. Biomed. Health Inform. 19 (3) (2015) 1146–1152.

[27] K. Matsunaga, A. Hamada, A. Minagawa, H. Koga, Image classification of melanoma, nevus and seborrheic keratosis by deep neural network ensemble, arXiv: 1703.03108.

[28] A. Krizhevsky, I. Sutskever, G.E. Hinton, Imagenet classification with deep convolutional neural networks, Advances in Neural Information Processing Systems, 25, Curran Associates, Inc., 2012, pp. 1097–1105.

[29] I. Lee, X. Du, B. Anthony, Hair segmentation using adaptive threshold from edge and branch length measures, Comput. Biol. Med. 89 (2017) 314–324.

[30] S. Chatterjee, D. Dey, S. Munshi, Mathematical morphology aided shape, texture and color feature extraction from skin lesion for identification of malignant melanoma, in: Proceedings of the IEEE CATCON 2015, December 2015, pp. 200–203.

[31] International Dermoscopy Society. <http://www.dermoscopy-ids.org>.

[32] Dermoscopy Atlas. <http://www.deroscopyatlas.com>.

[33] T. Mendonça, P.M. Ferreira, J. Marques, A.R.S. Marcal, J. Rozeira *PH2*—a dermoscopic image database for research and benchmarking, in: 35th International Conference of the IEEE Engineering in Medicine and Biology Society, July 3–7, 2013, Osaka, Japan.

[34] D. Gutman, et al., Skin lesion analysis toward melanoma detection: a challenge at the international symposium on biomedical imaging (ISBI) 2016, hosted by the international skin imaging collaboration(ISIC), 2016 [Online] Available: <https://arxiv.org/abs/1605.01397>.

[35] N. Dey, V. Rajanikanth, A.S. Ashour, J.M.R.S. Tavares, Social group optimization supported segmentation and evaluation of skin melanoma images, Symmetry 10 (2) (2018) 51.

[36] 32] Y. Yuan, Y. Lo, Improving dermoscopic image segmentation with enhanced convolutional–deconvolutional networks, IEEE J. Biomed. Health Inform. 23 (2) (2019) 519–526.

[37] V. Rajanikanth, S.C. Satapathy, N. Dey, S.L. Fernandes, K.S. Manic, Skin melanoma assessment using kapur's entropy and level set—a study with bat algorithm, Smart Intell. Comput Appl. 104 (2018) 193–202.

[38] M.A. Al-masni, M.A. Al-antari, M.-T. Choi, S.-M. Han, T.-S. Kim, Skin lesion segmentation in dermoscopy images via deep full resolution convolutional networks, Comput. Methods Prog. Biomed. 162 (2018) 221–231.

[39] J.L. Garcia-Arroyo, B. Garcia-Zapirain, Segmentation of skin lesions in dermoscopy images using fuzzy classification of pixels and histogram thresholding, Comput. Methods Prog. Biomed. 168 (2019) 11–19.

[40] L. Bi, K. Jinman, E. Ahn, D. Feng, Automatic skin lesion analysis using large-scale dermoscopy images and deep residual networks, 2017, arXiv:1703.04197.

[41] F. Navarro, M. Escudero-Vi∼nolo, J. Bescós, Accurate segmentation and registration of skin lesion images to evaluate lesion change, J. Biomed. Health Inform. 23 (2) (2019) 501–508.

[42] Y. Li, L. Shen, Skin lesion analysis towards melanoma detection using deep learning network, 2017, arXiv: 1703.00577.

CHAPTER 3

Extraction of effective hand crafted features from dermoscopic images

3.1 Introduction

For the development of computer-aided skin disease identification technique, implementation of efficient feature extraction algorithm is the most important stage. Quantification of significant information is the primary objective of the feature extraction algorithm. Considering the biomedical aspect, features explain the biological symptoms of the specific disease. For the skin disease identification, dermatologist consider various significant characteristics of skin lesions for the early stage detection of the disease. However, in-depth visualization of the affected area from dermoscopic images does not provide significant information for early stage detection of the disease. This qualitative analysis from visual inspection makes it very difficult to monitor the disease condition to a radiologist or expert dermatologist. The qualitative estimation of morphological properties and textural pattern is very much subjective in nature. Here, digital signal processing tools play a significant role to quantify the qualitative information from the skin lesion area. Development of signal processing tools for the extraction of specific attributes help to identify the disease at an early stage. Literature suggests various feature extraction techniques employing signal processing tools for the development of computer-aided skin disease identification. Barata et al. have used an image gradient-based histogram texture feature extraction technique along with the color feature extraction from six different color channels for the development of bag of features based melanoma detection system from dermoscopic images [1]. Rastgoo et al. have extracted completed local binary pattern, gray-level co-occurrence matrix (GLCM), histogram of oriented gradient, scale-invariant feature transform-based texture feature extraction techniques with shape feature descriptor for the differentiation of melanoma and dysplastic nevi [2]. Garnavi et al. have employed wavelet-based texture feature extraction technique with statistical morphological features and estimated border irregularity from the border series data for the computer-aided melanoma diagnosis system [3]. Kasmi and Mokrani [4] have proposed an automatic detection of ABCD features for the differentiation of benign lesions from melanoma. Employing the ABCD features, the melanoma and the benign lesions have been classified. Kawahara and Hamarneh have proposed a fully convolutional neural network architecture, interpolating feature maps from a number of intermediate layers of the

Recent Trends in Computer-aided Diagnostic Systems for Skin Diseases
DOI: https://doi.org/10.1016/B978-0-323-91211-2.00005-6

network for the detection of clinical dermoscopic features [5]. In Ref. [6], González-Díaz has proposed a computer-aided diagnostic system using convolutional neural network, incorporating the dermatologist's knowledge. In the proposed DermaKNet model an automatic dermoscopic structure segmentation and diagnosis module has been introduced, considering melanoma, nevus, and seborrheic keratosis (SK) diseases.

In this chapter, digital signal processing tools have been used to develop various feature extraction algorithms for the quantitative estimation of morphological, texture, and color features from dermoscopic images. Statistical morphological features have been extracted from segmented lesion area. Consequently, to quantify the irregular nature of the lesion area fractal dimension estimation technique has been employed. To incorporate finer border feature, wavelet—fractal-based border irregularity measurement technique is introduced. It appears to be a challenging task for the dermatologist and expert to differentiate closely similar skin abnormalities based on their textural variations from visual inspection of the lesion area. For the quantitative texture analysis fractal-based regional texture analysis (FRTA) and wavelet packet fractal texture analysis (WPFTA) algorithms are introduced. For the identification of skin lesions based on the presence of specific feature descriptor, annotated by the dermatologist, a cross-correlation-based technique is presented. using the cross-correlation technique spatial and spectral features have been extracted to identify melanocytic and nonmelanocytic skin lesions. In the subsequent sections of this chapter, different feature extraction algorithms along with the corresponding theoretical background have been elaborated.

3.2 Contemporary signal processing tools for feature extraction

Feature is synonymous to the explainable information of a particular attribute or variable. Properly representable features are considered to be an input to a classification problem. Extraction of relevant information from the associated signal helps to realize condition of the system in accordance to the expert's opinion. In case of biomedical signals the features represent the conditions of the associated biological systems monitored by doctors. Over the decades, signal processing tools have paved the leading path for extraction of meaningful features from the raw data. Signal processing tools have been explored to obtain significant features from biomedical signals or images for the development of automatic diagnostic algorithms. The following subsection portrays different signal processing tools used in this book for the extraction of relevant features from the biomedical images for further diagnosis of the diseases.

3.2.1 Fractal geometry and Hausdorff fractal dimension

Fractals generally show irregular and complex geometrical shapes that cannot be fully described by Euclidean dimension. Fractals describe a mathematical shape with a high degree of complexity. According to Mandelbrot [7], a set F is said to be fractal if it has a fine structure with high degree of irregularity. The fundamental consideration of fractal geometry is the notion of dimension [8,9]. Dimension indicates how much space a set occupies close to each of its points. Among the wide variety of "fractal dimensions," the definition of Hausdorff is the most important and convenient one. Hausdorff dimension is based on measures, which are relatively easy to manipulate. Mathematically, the dimension D_H of a geometrical set F is calculated by the following expression:

$$D_H(F) = \underset{d}{inf}\left\{\{d \geq 0 | K_H^d = 0\}\right\}$$

where D_H d-dimensional Hausdorff measure of F is defined by:

$$K_H^d(F) = \inf\left\{\sum_{i=1}^{\infty} |X_i|^d : \{X_i\} \text{ is a } r\text{—cover of } F\right\}$$

with $\{X_i\}$ specifying a finite collection of sets with diameter at most $r(r > 0)$, that covers F. An alternative and simpler form for the fractal dimension measurement of a complex object is given by,

$$D = \lim_{l \to 0} \frac{\log(M)}{\log(l)}$$

Here, l is the size of the geometrical shape, which has to be used M number of times to cover the fractal object. A general expression for the development of simple and efficient algorithm for the computation of fractal dimension of any fractal-like object represented in discrete domain is

$$D \propto \frac{\log(n(\delta))}{\log(\delta)}$$

where n is a measure of self-similarity and δ is a scale parameter.

3.2.2 Higuchi fractal dimension estimation

A computationally very fast algorithm for the direct estimation of the fractal dimension without estimation of the strange attractor has been proposed by Higuchi [10]. For a given signal $S(n)$, l new time series $S(n)_m^l$ are constructed as:

$$S(n)_m^l = s(m), s(m+p), \ldots, s\left(m + \left\lfloor \frac{N-m}{l} \right\rfloor \cdot l\right) \tag{3.1}$$

where, $m = 1, 2, \ldots, l$. Here, the initial time and the time interval have been indicated by m and l, respectively. The average length of m curves for each curve $S(n)$ has been obtained by:

$$L_{avg} = \frac{1}{l}\sum_{m=1}^{l}\left[\left(\sum_{i=1}^{\lfloor\frac{N-m}{l}\rfloor}|x(m+il)-x(m+(i-p)_p)|\right)\cdot\frac{N-1}{\lfloor\frac{N-m}{l}\rfloor p}\right] \quad (3.2)$$

where, $N-1/\lfloor\frac{N-m}{l}\rfloor p$ is a normalization factor. From the curve of $\log(L_{avg}(l))$ versus $\log(1/l)$, the slope of the least square linear best fit has been estimated as Higuchi's fractal dimension.

3.2.3 Katz fractal dimension estimation

Katz [11] proposed another methodology for the fractal dimension measurement of the time series waveform. The fractal dimension of a curve can be estimated in accordance with Eq. (3.3), where M can be considered as the total length or sum of the Euclidean distances between successive points and l as the diameter of the curve, considered as the maximum of distances between the first sample and all consequent samples in the time series. Katz proposed a normalization method for l and M by the length of the average step or average distance between successive points, s, defined as $s = M$, where n is the number of steps in the curve. Eq. (3.3) has been modified as:

$$D = \frac{\log(M/s)}{\log(l/s)} = \frac{\log(n)}{\log(l/M)+\log(n)} \quad (3.3)$$

Eq. (3.3) summarizes the Katz's methodology for fractal dimension measurement of a waveform.

3.2.4 2D wavelet packet decomposition

Although the Fourier transform provides significant information of a signal, wavelet transform extracts more information for signal analysis. Wavelet transform introduces small waves, called wavelets of varying frequency and limited duration. Introduction of wavelets of varying frequency and scales reveals not only the frequency information but also the time stamps. Therefore the wavelet transform is an efficient tool for time–frequency analysis or multiresolution analysis. In image processing problems, one-dimensional wavelet transform is extended to two-dimensional wavelet transform for the spatial–spectral analysis. Wavelet transforms use windows of varying width and offer greater flexibility for the analysis of images by detecting image discontinuities and details using short, high-frequency bases, and coarse features using longer, low frequency bases [12]. The wavelets measure intensity variations for an image along different directions. Therefore in two-dimensional wavelet transform, two-dimensional scaling function

($\varphi(p,q)$) along with three wavelet bases ($\psi^H(p,q), \psi^V(p,q)$, and $\psi^D(p,q)$) as a product of two one-dimensional functions, are considered. The two-dimensional scaling function and the wavelet bases can be expressed as follows:

$$\varphi(p,q) = \varphi(p)\varphi(q) \tag{3.4}$$

$$\psi^H(p,q) = \psi(p)\varphi(q) \tag{3.5}$$

$$\psi^V(p,q) = \varphi(p)\psi(q) \tag{3.6}$$

$$\psi^D(p,q) = \psi(p)\psi(q) \tag{3.7}$$

Here, these wavelets estimate the intensity variations of the image along three different directions. While ψ^H provides variations along horizontal edges (along columns), ψ^V gives variations along vertical edges of the image (along rows) and ψ^D determines intensity variations along diagonals. Therefore the two-dimensional scaling function and wavelet bases can be defined in the following manner:

$$\varphi_{j,r,c}(p,q) = 2^{(j/2)}\varphi(2^j p - r, 2^j q - c) \tag{3.8}$$

$$\psi^i_{j,r,c}(p,q) = 2^{(j/2)}\psi^i(2^j p - r, 2^j q - c), i = H, V, D \tag{3.9}$$

where i indicates the directional wavelets according to the above equations. Considering these two-dimensional scaling function and wavelet bases, the two-dimensional wavelet transform of an image $I(p, q)$ of size $P \times Q$, can be defined as an extension of one-dimensional wavelet transform as:

$$W_\varphi(j_0,r,c) = \frac{1}{\sqrt{PQ}}\sum_{p=0}^{P-1}\sum_{q=0}^{Q-1} I(p,q)\varphi_{j_0,r,c}(p,q) \tag{3.10}$$

$$W^i_\psi(j,r,c) = \frac{1}{\sqrt{PQ}}\sum_{p=0}^{P-1}\sum_{q=0}^{Q-1} I(p.q)\psi^i_{j,r,c}(p,q), i = H, V, D \tag{3.11}$$

Similar to the one-dimensional wavelet transform, $W_\varphi(j_0,r,c)$ defines the approximate coefficients of $I(p, q)$ at scale j_0. $W^i_\psi(j,r,c)$ corresponds to the horizontal, vertical and diagonal details or directional detailed coefficients for scales $j \geq j_0$. Here, $j_0 = 0$ and $P = Q = 2^j$ such that $j = 0, 1, 2, \ldots J - 1$ and $r = c = 0, 1, 2, \ldots 2^j - 1$. Wavelet transform decomposes an image into four parts: an approximation component and three detailed components. The horizontal component measures variations along horizontal edges, the vertical component responds to variation along rows or vertical edges and the diagonal component is related to variations along the diagonals. Schematic representation of wavelet decomposition is shown in Fig. 3.1. The wavelet packet decomposition (WPD), originally known as optimal subband tree structuring, offers more detailed description of scaling behavior into position, frequency, and scale terms than the discrete wavelet transform by iteratively decomposing the image detailed components. This results in a more flexible multiresolution analysis. In WPD, both the approximations and the detailed components are further decomposed into four coefficients to create a complete binary tree. The parent node is decomposed into one approximate and three detailed components. In the second stage, each of all those previous coefficients is further decomposed into a similar manner into four coefficients to generate 16 coefficients. This process is continued until no further decomposition is possible. As shown in Fig. 3.2, N_0 is the original image or parent node, and it has been decomposed into approximate and detailed coefficients (horizontal, vertical and diagonal) denoted as N_{A1}, N_{H1}, N_{V1}, and N_{D1}, respectively, in level 1. Similarly at level 2, the horizontal component has been further decomposed as $N_{H1.1}$, $N_{H1.2}$, $N_{H1.3}$, and $N_{H1.4}$, and so on.

3.2.4.1 Wavelet transforms using Matlab

The function *dwt2* is use for single-level 2D discrete wavelet transform of an image. This function decomposes an input image into approximate and detail coefficients according to the selected "mother wavelet."

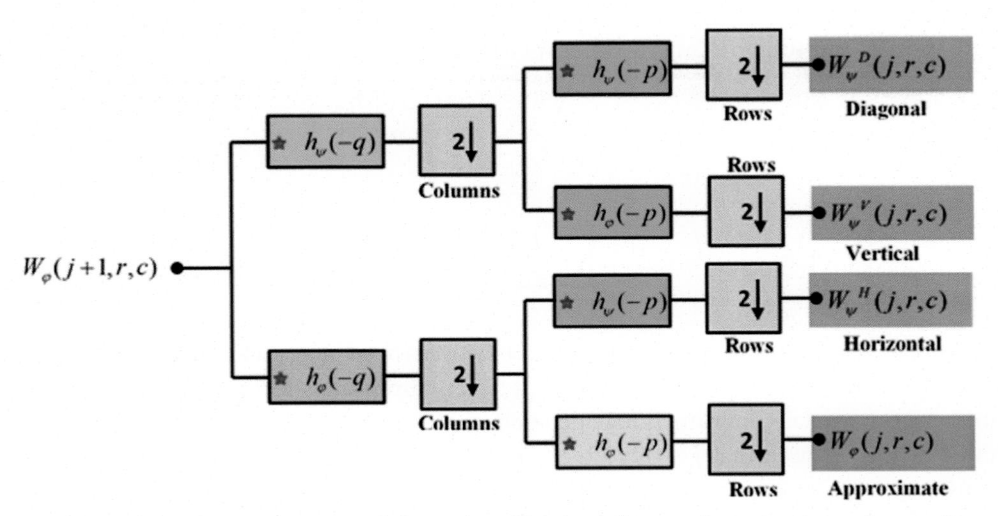

Figure 3.1 Schematic representation of the analysis filter bank for two-dimensional wavelet transform.

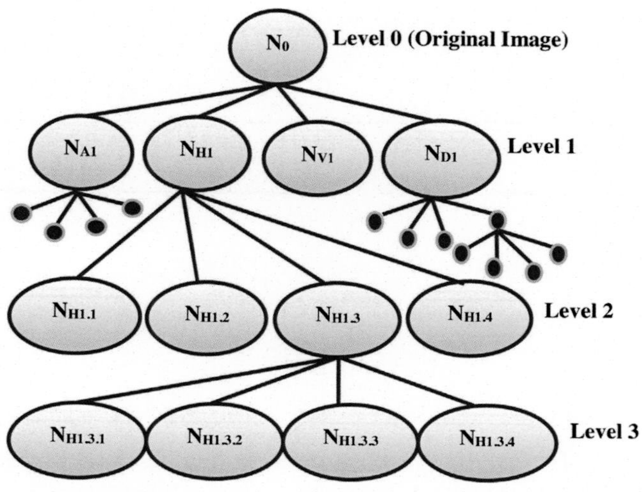

Figure 3.2 Schematic representation of the wavelet packet decomposition technique.

Syntax:

$$[W_A, W_H, W_V, W_D] = dwt2(I, \text{"}mwavelet\text{"})$$

dwt2 computes the single-level 2D wavelet decomposition of the input image I. Here, W_A, W_H, W_V, W_D corresponds to the approximates coefficient, horizontal coefficient, vertical coefficient, and diagonal coefficients, respectively. *"mwavelet"* is the analyzing mother wavelet to compute 2D discrete wavelet transformation. The mother wavelet belongs to the following wavelet families: Daubechies, Coiflets, Symlets, Fejér-Korovkin, Discrete Meyer, Biorthogonal, and Reverse Biorthogonal.

Name of the wavelets of different orders are given as:

Daubechies—"db1," "db2," . . ., "db10," . . ., "db45"
Coieflets—"coif1," "coif2," . . ., "coif5"
Symlet—"sym2," "sym2," . . ., "sym45"

The function *idwt2* is used for inverse 2D wavelet transform. The inverse wavelet transform is employed to reconstruct the original image from the corresponding approximate and detail coefficients W_A, W_H, W_V, and W_D.

Example:

I = *rgb2gray* (*imresize* (*imread* (*"file name"*), [*512,512*])); - read the original image, resize it to 512 × 512 in size and convert it into gray scale.

imshow(I)—display the original grayscale image

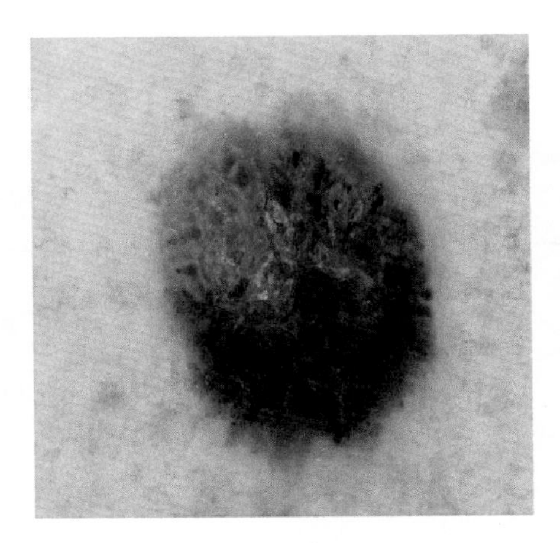

$[W_a, W_h, W_v, W_d]$ = *dwt2(I, "db3");*—perform 2D discrete wavelet transform of the image *I* with 3rd order Daubechies mother wavelet "db3."

figure, imagesc(W_a)

title("Approximate Coefficient")—display the Approximate coefficient

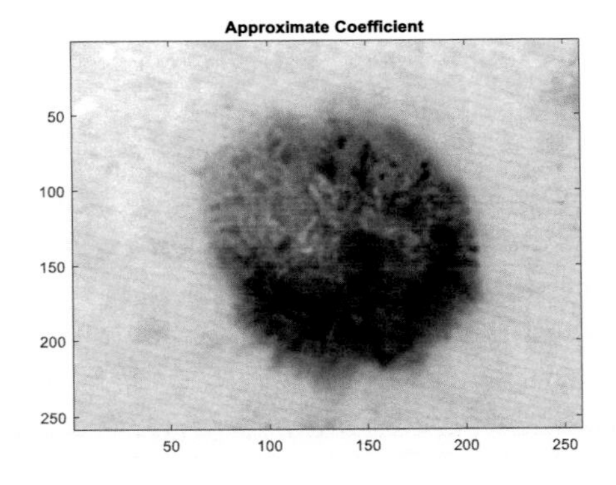

figure, imagesc(W_v)

title("Approximate Coefficient")—display the detail Horizontal coefficient

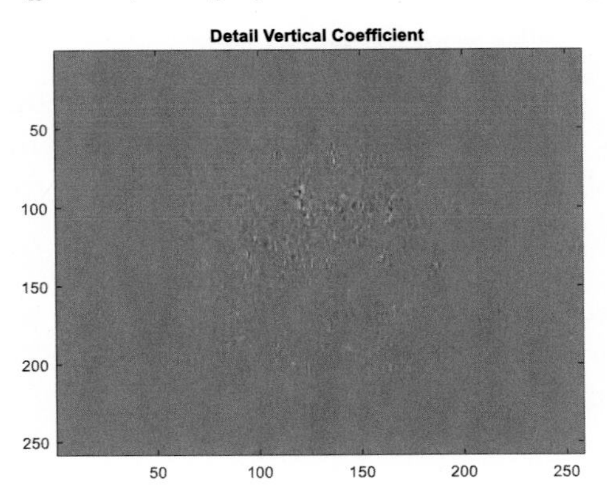

Syntax:

$$I_r = idwt2(W_A, W_H, W_V, W_D, \text{"}mwavelet\text{"})$$

idwt2 reconstruct the image I_r from the approximate and detail coefficient matrices.

The function *wpdec2* performs the WPD of input image.

Syntax:

$$I_{PD} = wpdec2(I, L, \text{"}mwavelet\text{"})$$

wpdec2 decomposes the input image I into L decomposition levels with a specified mother wavelet. By L level decomposition N number of nodes are obtained. The coefficient values of the corresponding node can be computed using the following Matlab function:

$$W_N = wpcoef(I_{PD}, \text{"N"})$$

3.2.5 Empirical wavelet transform: a cursory view

Wavelet analysis is one of the most efficient tools for signal and image processing. For multiresolution analysis, a mother wavelet of varying width with low and high-frequency bases is used for detecting the image discontinuities. A particular rigid mother wavelet is used to construct a filter bank for the time–frequency analysis of an image. An adaptive extension is employed in the WPD technique by successive refinement of the scale. However, a constant ratio in the successive subdivision limits the adaptability [13,14].

Empirical wavelet transform (EWT) is an adaptive methodology for spatial–spectral analysis of an image to construct wavelet bases directly from the information contained in the image itself. The EWT consists of two major steps: estimation of the Fourier supports followed by construction of corresponding wavelet bases as per those supports. In EWT, considering the Shannon criterion, the normalized Fourier support from 0 to π is segmented into S contiguous segments, $\omega_0 = 0$ to $\omega_S = \pi$. Each segment is centered around ω_s with a transition phase of T_s of width $2\tau_s$. The arbitrary function $\beta(m)$ can be defined as [14]:

$$\beta(m) = \begin{cases} 1 & \text{if } m \leq 0 \\ \text{and } \beta(m) + \beta(1-m) = 1 & \forall m \in [0,1] \\ 1 & \text{if } m \geq 1 \end{cases} \tag{3.12}$$

Considering τ_s proportional to ω_s:$\tau_s = \lambda\omega_s$ where $0 < \lambda < 1$ empirical scaling function ($\xi_m(\omega)$) and the empirical wavelets ($\varsigma_m(\omega)$) are defined as band pass filters and expressed as the following:

$$\xi_m(m) = \begin{cases} 1 & \text{if } |\omega| \leq (1-\lambda)\omega_m \\ \cos\left[\frac{\pi}{2}\beta\left(\frac{1}{2\lambda\omega_s}(|\omega| - (1-\lambda)\omega_s)\right)\right] & \\ \quad \text{if } (1-\lambda)\omega_s \leq |\omega_s| \leq (1+\lambda)\omega_s & \\ 0 & \text{otherwise} \end{cases} \tag{3.13}$$

$$\varsigma_m(\omega) = \begin{cases} 1 & \text{if } (1+\lambda)\omega_m \leq |\omega| \leq (1-\lambda)\omega_m \\ \cos\left[\frac{\pi}{2}\beta\left(\frac{1}{2\lambda\omega_{m+1}}(|\omega| - (1-\lambda)\omega_{m+1})\right)\right] & \\ \quad \text{if}(1-\lambda)\omega_{m+1} \leq |\omega| \leq (1+\lambda)\omega_{m+1} & \\ \sin\left[\frac{\pi}{2}\beta\left(\frac{1}{2\lambda\omega_m}(|\omega| - (1-\lambda)\omega_m)\right)\right] & \\ \quad \text{if}(1-\lambda)\omega_m \leq |\omega| \leq (1+\lambda)\omega_m & \\ 0 & \text{otherwise} \end{cases} \tag{3.14}$$

In this proposed technique, 2D empirical Littlewood–Paley wavelet transform has been employed to decompose the input image in low- and high-frequency components. To perform 2D EWT of the image (I), 1D Fourier transform is performed along each row and column of the image to calculate the mean row and column magnitude spectrum. Fourier boundaries have been detected from row and column spectrums to construct the corresponding filter bank. From each of the filter, corresponding subband images are obtained.

3.2.6 Sparse autoencoder

Unsupervised learning algorithms are able to learn discriminative and effective features from a large amount of unlabeled dataset. An autoencoder, a symmetrical neural network learns features in an unsupervised manner to produce a novel useful representation of those features using backpropagation algorithm. The autoencoder tries to learn an approximation to the identity function, such that the output $\widetilde{\mathcal{X}}$ is similar to the input $\mathcal{X}$. Learning the approximation function $\langle_{W,b}(\mathcal{X})$ in the hidden layer for the input feature matrix $\mathcal{X}$ the autoencoder obtain a feature expression $\langle(\mathcal{X}^{(i)}, W, b) = \sigma(W\mathcal{X}^{(i)} + b)$, where $i = 1, 2 \ldots, N$ at the hidden layer to ensure the output $\sigma(W^T\langle(\mathcal{X}^{(i)}, W, b) + c)$ close to the input. The average activation of a hidden layer i can be defined as [15]:

$$\beta_i = \frac{1}{n}\sum_{i=1}^{n}[a_j(x(i))] \tag{15}$$

To ensure the average activation of each hidden neuron β_i is mostly inactive, a sparse term has been added to the objective function as a penalty term, expressed as the following:

$$\mathcal{P}_{\surd} = \sum_{i=1}^{M} KL(\beta || \beta_i) \tag{3.16}$$

where M is the number of neurons in a hidden layer and Kullback–Leibler divergence ($KL(.)$)can be expressed as:

$$KL(\beta || \beta_i) = \beta \log \frac{\beta}{\beta_i} + (1 - \beta) \log \frac{1 - \beta}{1 - \beta_j} \tag{3.17}$$

Therefore the cost function of the neural network considering the sparse penalty term can be given as:

$$C_{f_{\surd}} = \left[\frac{1}{n} \sum_{i=1}^{n} \left(\frac{1}{2} \| \langle_{w,b}(x(i)) - y(i) \|^2 \right) \right] + \frac{\delta}{2} \sum_{l=1}^{k-1} \sum_{i=1}^{M_l} \sum_{j=1}^{M_{l+1}} (W_{i,j}(l)) + \alpha \sum_{i=1}^{M} KL(\beta || \beta_i) \tag{3.18}$$

where α has been considered as the weight of the sparsity penalty.

3.2.7 Cross-correlation technique

The cross-correlation of two finite-duration 1D sequences enables to find the resemblance of the sequences, when one sequence is shifted relative to the other, along the abscissa [16–18]. The cross-correlation ($\Gamma_{f,g}$) is given by:

$$\Gamma_{f,g}(n) = \sum_{m=-\infty}^{\infty} f(m) g \times (m - n) \tag{3.19}$$

where f and g are the two sequences, each of length m samples, and n is the lag parameter.

Cross-correlation of an image with a feature descriptor (a kernel image of smaller size) is obtained as the descriptor scans across the larger image, and helps to find out the section of the image that fits to the selected feature descriptor. The cross-correlation of an input image (I) of size ($P \times Q$) with a template or kernel image ($\mathbf{K}$) smaller in size ($l \times m$) can be expressed as:

$$\mathbf{I_C}(s, t) = \sum_{p=0}^{P-1} \sum_{q=0}^{Q-1} \mathbf{I}(p, q) \mathbf{K} \times (p - s, q - t), \quad \begin{array}{l} -(l-1) \le s \le P - 1, \\ -(m-1) \le t \le Q - 1, \end{array} \tag{3.20}$$

Symbols s and t represent the row and the column indices of the resultant cross-correlogram ($\mathbf{I_C}$). In Eq. (3.20), the negative and positive row indices correspond to upward and downward shift of the rows of the kernel **K**, respectively. Similarly, the leftward and the rightward shift of the columns have been denoted by negative and positive column indices of kernel **K** [17]. The cross-correlation operation has been used to find the presence or absence of a specific feature descriptor in an input image. The correlation coefficient matrix $\gamma(r, c)$ has been obtained by normalizing the cross-correlated output matrix by the local sum using the following equation [16]:

$$\gamma(r, c) = \frac{\sum_{s,t}[I(s,t) - \overline{I}_{r,c}][K(s-r, t-c) - \overline{K}]}{\left\{\sum_{s,t} I(s,t) - \overline{I}_{r,c}{}^{2} \sum_{s,t}[K(s-r, t-c) - \overline{K}]^{2}\right\}^{0.5}} \tag{3.21}$$

3.2.7.1 Cross-correlation using Matlab

The Matlab function *xcorr2* computes the cross-correlation of two matrices or two images or an image with a template.

Syntax:

$$X_C = xcorr2(I, K)$$

xcorr2 returns a matrix X_C after the cross-correlation operation of two matrices I and K without scaling. The autocorrelation matrix of input I can be evaluated by using the same function as *xcorr2(I, I)*.

For the matrix I of size $M \times N$ and K of size $P \times Q$, their 2D cross-correlation output matrix X_C have a size of $(M + P - 1) \times (N + Q - 1)$.

To obtain normalized cross-correlogram the following function is used:

$$X_C = normxcorr2(I, K)$$

3.3 Extraction of morphological features from skin lesions

Structural properties of skin lesions are important for the realization of the shape of the affected area and further spreading of the disease. Dermatologists follow the morphological features to estimate the structural complexity and irregularity.

3.3.1 Statistical morphological features

This section illustrates the quantitative assessment of the features related to the shape of the lesions, such as area, perimeter, equivalent diameter, and so on. These shape related features are estimated as follows:

- *Area*: Number of pixels present in segmented lesion area.

- *Perimeter*: Number of pixels along the border of the lesion.
- *Major axis length*: Length (in pixels) of the major axis of the ellipse that has the same normalized second central moments as the region, returned as a scalar.
- *Minor axis length*: Length (in pixels) of the minor axis of the ellipse that has the same normalized second central moments as the region, returned as a scalar.
- *Equivalent diameter* (ED): Diameter of a circle with the same area as the region:

$$\mathrm{ED} = \sqrt{\frac{4 \times \mathrm{Area}}{\pi}}$$

- *Circularity*: It specifies the roundness of objects. The circularity value is computed as:

$$\mathrm{Circularity} = \frac{4 \times \mathrm{Area} \times \pi}{(\mathrm{Perimeter})^2}$$

- *Rectangularity*: The ratio of the object area and the area of the smallest rectangle containing the region, that is, the object-oriented bounding box.
- *Elongation*: The ratio of the height and the width of the smallest rectangle containing the object.
- *Aspect ratio*: The ratio of the length of major axis to the length of minor axis of the segmented object.
- *Solidity*: A measure of border irregularity defined as the ratio of the object and the smallest convex polygon that can contain the region.
- *Eccentricity*: It is the ratio of the distance between the foci of the ellipse, having same second-moment as the region and its major axis length.
- *Average distance*: Mean of all the measured Euclidean distances between the center pixel and all border pixel locations.
- *Distance variance*: Variance of all the measured Euclidean distances between the center pixel and all border pixel locations.

3.3.2 Application of techniques for measurement of lesion-border irregularity

Irregularity of the skin lesion is another critical criterion for the diagnosis of malignant melanoma. In the existing literature, most of the authors have determined the border irregularity by estimating the distance variance of border pixels to the center pixel. In this section, a border irregularity measurement algorithm has been discussed by measuring the fractal dimension of the segmented object to quantify the irregular nature of the skin lesion.

3.3.2.1 Fractal dimension estimation for border irregularity measurement

- *Materials*: In this study, primarily 6579 digital dermoscopic image consisting of 2419 melanoma cases, 3261 dysplastic nevi cases and remaining 899 subjects with basal cell carcinoma (BCC) have been composed from the dataset available over the Internet [19–22]. The expert dermatologists have histopathologically confirmed the diseases associated with the images considered for analysis.
- *Methodology*: Fractals generally show irregular and complex geometrical shapes that cannot be fully described by Euclidean dimension. Fractals describe a mathematical shape with a high degree of complexity. According to Mandelbrot [7], a set F is said to be fractal if it has a fine structure with high degree of irregularity. In this study, the Haussdorf fractal dimension [9] of the skin lesion border has been measured to assess the irregularity of the complex geometrical shape of the lesion. The log–log curve has been obtained by considering the logarithm of scaling factor, that is, the inverse of box size (1/box size) along the x-axis and the logarithm of number of boxes required to fill the region along the y-axis. The slope of the curve has been considered as the fractal dimension of that border image.
- *Results and discussions*: For the estimation of the border irregularity, the segmented image is processed in the border detection algorithm as explained in the previous chapter. From the border detection algorithm, single pixel border of the lesion area is obtained, that illustrates the irregular nature of the affected area. The fractality curves for the fractal dimension measurement of corresponding border detected images have been shown in Fig. 3.3. Fractality curves have been shown for various skin lesions having different order of structural complexity. The high value of fractal dimension corresponds to the presence of more irregularity of the object. In Fig. 3.3, the fractality curves for melanoma, nevus, and BCC images have been shown. From the estimated fractal dimension, it has been observed that melanoma lesion has more irregular structure than nevus and BCC.

Some of the lesion borders contain different notches that may not be encountered during the calculation of Higuchi fractal dimension (HFD) using box counting method. To address this issue, a one-dimensional border series have been constructed by considering distance variations of each border pixels from the center pixel location. Using this border series, the abrupt changes along the border region has been determined for proper irregularity estimation.

3.3.2.2 Wavelet–fractal-based border irregularity measurement

- *Materials*: The entire study has been performed on the 4094 skin lesion images of two different classes, namely common melanoma and benign nevi diseases. The skin lesion images have been collected from some widely acceptable image databases [19–22].

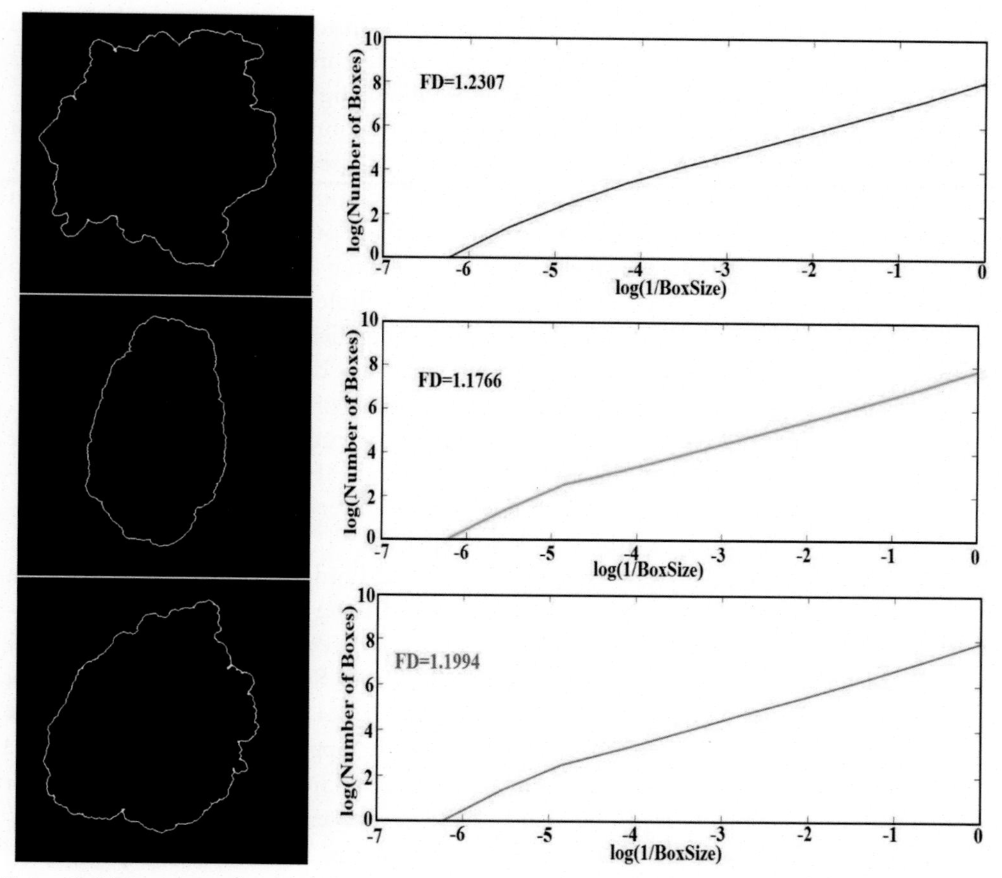

Figure 3.3 The curves for the fractal dimension measurement of the corresponding border images. Fractal dimension (FD) denotes the border irregularity present in the lesion.

- *Methodology*: The irregular nature of the skin lesion is a very important and distinguishable feature for the proper identification of skin abnormalities. Therefore from single-valued fractal dimension, proper estimation of irregularity with smaller notches is very much challenging. Inspired by the work on boundary-series features by Garnavi et al. [3], a fractal-based border irregularity measurement technique in wavelet domain has been reported here. In this study, the distance between the center pixel location and each of the border pixel location has been calculated following the equation:

$$d_i = \sqrt{(x_i - x)^2 + (y_i - y)^2} \tag{3.22}$$

where d_i the measured distance, (x_i, y_i) and (x, y) denote the border pixel location and the center pixel location respectively. A border series has been obtained by constructing a finite length data as the pixel indices considered along the x-axis and the measured distance along the y-axis. To estimate the irregularity with more finer details, the border series data has been decomposed into high and low frequency components. The high-frequency components depicted the abrupt changes in distance of the border pixels from the center pixel while the low frequency component correspond to the smooth variations. A three-level WPD technique has been used to obtain the approximate and the detailed components of the border series data. From this three-level pyramid structured wavelet transforms, three pairs of approximate and detailed components have been obtained. On the whole seven border series have been obtained by reconstructing all the decomposed coefficients, along with the original border series. Each of the obtained border series has been analyzed by means of extracting some histogram features namely mean, variance, skewness, kurtosis, energy, and entropy. Other than the statistical features, the irregular nature of each of the obtained border series has been quantified by measuring HFD and Katz fractal dimension. The above mentioned fractal dimension estimation technique quantify the anomalous nature along the border of the skin lesion area.

- *Results and discussions*: In this wavelet—fractal-based border irregularity measurement technique, the border series have been decomposed using lower order mother wavelets, namely Daubechies, Symlets, and Coiflets. For WPD, 1st-order Daubechies to 10th-order Daubechies, 2nd-order Symlet to 8th-order Symlet and 1st-order Coiflet to 5th-order Coiflet 5 filters have been chosen. In Fig. 3.4 the original border series of a lesion-border image has been shown along with the approximate as well as detailed representation of first level wavelet decomposition using 3rd-order Daubechies mother wavelet. From the figure, it can be seen that the border series depicts the proper irregularity of the lesion area by means of distance variations. Therefore the estimation of fractal dimension from border series helps to determine appropriate structural irregularity. Further decomposition of the border series portrays more detailed descriptor of border irregularity. Estimation of fractal dimension from the reconstructed border series from wavelet coefficients ensures the fractal measurement of finer details.

3.4 Application of texture feature extraction techniques

Among the wide varieties of features, texture feature is the most subjective in nature. It is very challenging for the dermatologist and expert to estimate the textural pattern along the skin lesion area from visual inspection. The textural complexity of skin lesion varies according to the wide varieties of skin diseases. The intensity variation along the lesion area has been quantified using various texture feature extraction algorithms. In this section, GLCM technique is employed to estimate the intensity variation of the skin lesions. FRTA and wavelet—fractal-based texture analysis techniques are introduced to quantify regional and finer textural variations.

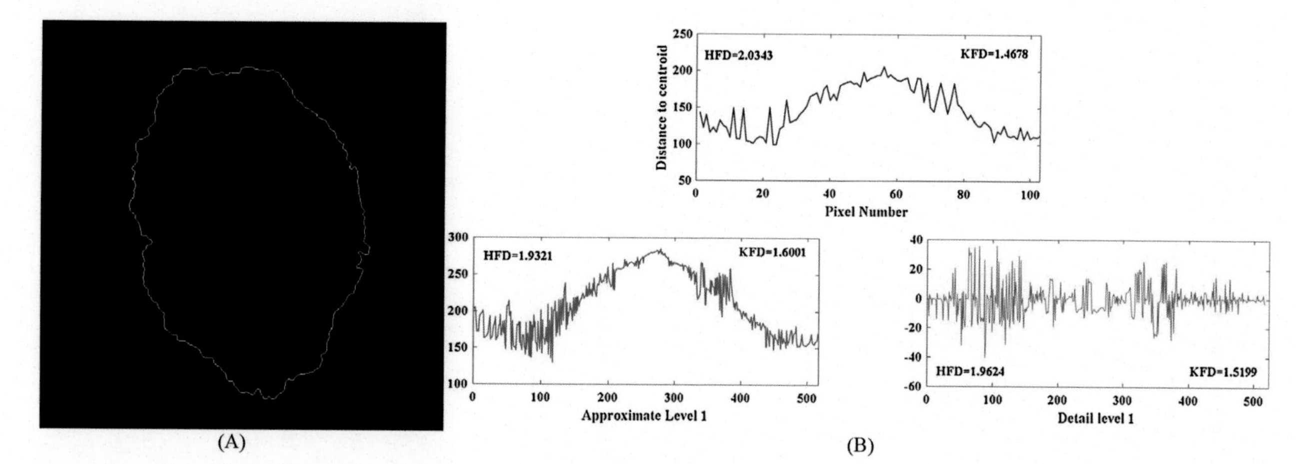

Figure 3.4 (A) Border detected image of a benign nevi, (B) original border series (top) an approximate and detail representation at level 1 wavelet decomposition using Daubechies mother wavelet (bottom). *HFD*, Higuchi fractal dimension; *KFD*, Katz fractal dimension.

3.4.1 Gray-level co-occurrence matrix-based texture analysis

3.4.1.1 Materials

The composite image database has been formed in a manner similar to what has been done in connection with fractal dimension estimation for border irregularity measurement.

3.4.1.2 Methodology

GLCM function characterizes the texture of an image by means of calculating how often a pair of pixels with specified values and spatial relationship occurs in an image. To construct the GLCMs, the parameters have been selected through rigorous analysis for the extraction of most important textural information with lesser number of correlated features. GLCMs have been constructed by considering the quantization levels as $N=8$, 16, 32 and keeping constant pixel distance (9 pixel distances) and the directions of rotation from 0 degree to 135 degree with 45 degree increment from the major axis. Finally, the statistical features have been computed from the GLCMs obtained with the following settings: (16 quantization levels, that is, $N=16$, nine pixel distances and four directions $\theta=0$ degree−135 degree from the major axis) to extract the most efficient representative features. From each GLCM, the statistical features are extracted from each of the original grayscale dermoscopic images, as correlation, energy, homogeneity, contrast, autocorrelation, dissimilarity, entropy, and maximum probability, sum-squared variance, sum average, sum variance, sum entropy, difference variance, and difference entropy.

3.4.2 Fractal-based regional texture analysis

3.4.2.1 Materials

In this study, the same dataset has been used as in GLCM-based texture feature extraction technique.

3.4.2.2 Methodology

Fractal dimension analysis is used to describe ruggedness, complexity, roughness, or irregularity of an object depending on the set to which it belongs. Texture can be considered as a repetition of similar pattern distributed on the entire region of the object. Costa et al. have used segmentation-based fractal texture analysis for the extraction of textural information from the grayscale images in Ref. [23]. A set of binary images have been obtained from the input grayscale image using the proposed algorithm. In order to describe segmented texture patterns the fractal dimensions of the resultant regions are computed from each of the binary images. In this study, a FRTA technique has been introduced. In this FRTA algorithm, number of decomposed regions has been optimized over the region of interest by estimating the local intensity variation and fractal dimension. This reported FRTA algorithm has introduced a texture analysis technique by quantifying the intensity distribution and textural complexity present in the smaller subregions of the entire image. Application of multilevel Otsu's method for the decomposition of the dermoscopic image into binary images has sometimes led to oversegmentation and unnecessary decomposition of the background region of the image. Considering the segmented binary image as a binary mask, the region of interest has been extracted from the original grayscale image. It restricts the oversegmentation and also ensures further decomposition of the grayscale regions from the skin lesion area only, irrespective of the type of surrounding and the skin textural pattern of the patient. In the FRTA algorithm, the decomposition has been initiated by initializing the decomposition level as the minimum possible gray levels of the image. The masked original grayscale image with initial minimum possible gray levels of the image (n_d) has been considered as an input parameter of the algorithm. Multilevel Otsu's algorithm [23] has been implemented for the generation of binary images from the input image. The multilevel Otsu's algorithm determines the threshold value that minimizes the input image intraclass variance considering the gray-level distribution of the image, and applies it to each image region until the desired number of threshold values (n_d) have been obtained. After obtaining the set of threshold values according to the predefined parameter, the grayscale dermoscopic image has been disintegrated into a collection of binary images using two separate approaches. In the first approach, the image has been segmented using the set of threshold values individually; that is, the number of binary images is same as the predefined parameter value. In the second approach, a pair of thresholds (lower and upper) have been selected and applied to original grayscale image to get the set of binary images.

$$I_b(m,n) = \begin{cases} 1 & \text{if } T_l < I(m,n) < T_h \\ 0, & \text{otherwise.} \end{cases} \tag{3.23}$$

where I_b is the resultant binary image and T_l and T_h are the selected lower and upper threshold values for the original image. From each of the decomposed regions of the image, after estimating the mean intensity and fractal dimension value, the grayscale region has been further decomposed. As in Otsu's method, the region boundary from each of the resultant binary images has been extracted using eight-connected neighbor properties as follows:

$$P_b(i,j) = \begin{cases} 1 \textit{ if } \exists(i',j') \in N_8[(i,j)] \\ \quad P_t(i',j') = 0\Lambda \\ \quad P_t(i',j') = 1, \\ \quad 0, \text{ otherwise} \end{cases} \tag{3.24}$$

where $P_b(I, j)$ has been considered as the border image and the eight-connected pixels to (i, j) has been denoted by $N_8[(i, j)]$, [24]. The FRTA algorithm has been elaborated in Algorithm 1. To construct the FRTA feature vector, the area, the mean gray level, and the boundaries' fractal dimension have been measured. As mentioned previously, the fractal dimension has described the complexities of the boundaries of regions and segmented structures. The box counting algorithm has been used to compute the fractal dimension from each of the image segments [24–28]. Regional textural complexity has been extracted from the fractal features of each of the smaller subregions of the skin lesions.

Algorithm 1:
FRTA feature extraction

Input: Grayscale image I, initial number of thresholds $n_d = 2$
Output: Feature vector FV_{FRTA}
1: $T \leftarrow Multilevel\ Otsu's\ Method(I, n_d)$
2: $T_1 \leftarrow \mathrm{t}_1, \mathrm{t}_2 \in T$
3: $T2 \leftarrow \{t_1, t_2, t_{max}\}; t_{max} \rightarrow maximum\ Intensity(I)$
4: $i \leftarrow 1$;
5: $for\ \ \{\{t_i\}:\{t_i\} \in T_1\}\ \ do$
6: $I_b \leftarrow Binary\ Thresholding(I, t_i)$
7: $I_{border} \leftarrow Find\ Borders(I_b)$
8: $FV_{FRTA} \leftarrow [Box\ Counting(I_{border}), mean\ Gray\ Level(I, I_b), Fractal\ Feature(I, I_b)]$
9: $end\ \ for$
10: $i \leftarrow 1$;
11: $for\ t_l \leftarrow t_i; t_u \leftarrow t_{i+1}:\{t_i\} \in T_2; i \in [1 \cdots |T_2| - 1]$
12: $Repeat\ \ Step\ 6\ to\ 8$
13: $end\ \ for$
14: $Repeat\ \ Step\ 1\ to\ 13\ for\ each(I, I_b)$

3.4.2.3 Results and discussions

For each of the dermoscopic images, multilevel Otsu's method determines the appropriate threshold values for segmentation of various subregions. The number of threshold value is initialized as 2. Therefore the pixels belong to these two threshold values have been segregated. On the other hand, considering the maximum intensity of the lesion area, the consecutive values have been chosen as lower and upper threshold values for the segregation of subregions. Therefore using this technique three subregions have been segmented in the first iteration. Each subregions have been segmented again by employing multilevel Otsu's method. This technique disintegrates the entire lesion area into smaller subregions until minimum interclass variance is reached. Fig. 3.5 depicts some significant textural regions of the original grayscale images of melanoma, nevus, and BCC lesions. From the figure it has been found that FRTA algorithm has segregated various regions from the lesion area according to the regional intensity variations. Estimation of regional features and fractal dimension determines the regional textural complexity of the lesion area.

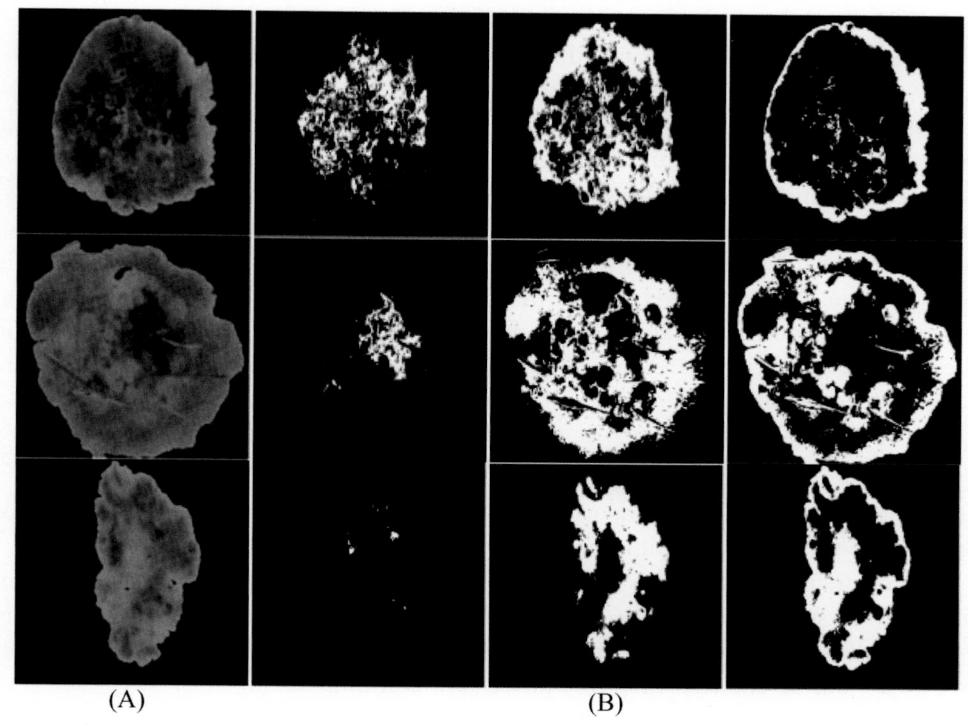

Figure 3.5 (A) Original grayscale image masked with the segmented image; (B) extracted subregions using FRTA algorithm and corresponding fractal curve.

3.4.3 Wavelet packet fractal texture analysis

Understanding the visual texture plays an important role in computer-aided diagnosis of skin diseases. As reported in the existing literatures, the texture analyses are performed either in spatial domain, or in frequency domain, or even using the combination of both. In the spatial domain, the texture features are extracted by means of some statistical measures of local texture patterns, whereas frequency domain analyses give some other essential characteristics of texture. To use the spatial and frequency domain information, some authors have proposed spatial-spectral domain approach based on wavelet transform for the texture analysis. Here, a texture analysis technique is illustrated that combines the wavelet-based representation of the dermoscopic images and fractal texture analysis method, to obtain detailed and robust texture information.

3.4.3.1 Materials

The entire study has been done on the 4094 skin lesion images of two different classes, namely common melanoma and benign nevi diseases. The skin lesion images have been collected from some widely acceptable image databases [19–22].

3.4.3.2 Methodology

The estimated single-valued fractal dimension does not adequately represent the complexity of the object. The entire fractality curve can however serve as a powerful representative model of the region of interest. This is known as fractal descriptor of an object. The fractality values under different scales provide complete information of the distribution of image pixel intensity across the image region. This important information on visual or physical attributes of the image region of interest leads to more precise and robust methodology to describe complex nature of the object. As discussed in Ref. [29], the wavelet packet transform gives a large number of wavelet coefficients denoted as D^j_i, where i is the scale component and j is the translation component. The wavelet coefficients are sorted in descending order according to their discrete energy value as follows:

$$\overline{D}:\{|D_1| \geq |D_2| \geq |D_3| \geq \ldots |D_m| > 0\} \tag{3.25}$$

Here, the zero valued components of the wavelet coefficient vector have been discarded since they are not significant for the power-law scaling, as described below:

The coefficient vector $\overline{D}$ is very large in size and most of its coefficients are contaminated by statistical noise. So, to consider the information containing coefficients, only the elements of the $\overline{D}$ vector at exponential intervals of $n = 1, 2, 4, 8, \ldots 2^{\log_{2m}}$ have been considered, according to a power-law relation.

The resultant vector in descending order is given by:

$$D:\{\overline{D}(1), \overline{D}(2), \overline{D}(4), \overline{D}(8), \ldots \overline{D}(2^{\log_{2m}})\}, \tag{3.26}$$

D scales with n according to a power-law relation:

$$D \propto n^{\lambda_D}$$

The procedure described above can be used for the estimation of the fractal descriptor of an image containing fractal-like objects, as this dimension is linearly related to λ_D.

The skin lesion images have self-similar nature with a uniform intensity distribution in the normal skin region, whereas a noticeable intensity variations along the region of interest. This textural pattern makes the texture analysis of skin lesion image more subjective in nature. The characteristics, color, and pattern of skins of different people vary across the different regions of the world, and also according to the age. The lesion may occur in various locations of human body which have skins of different nature. As an example, the skin of human skull differs from the skins of other regions of human body. As the textural pattern of the lesion area is the subject of interest, the texture analysis of the entire dermoscopic image may not lead to the extraction of proper textural information. To overcome this problem, the segmented image (the region of interest) has been masked with the original color dermoscopic image for the exclusion of the normal skin areas from the purview of textural analysis. The wavelet transform of two-dimensional images give more detailed information about the intensity distribution of the digital images. The texture analysis in spatial domain based on different statistical features is not sufficiently describing the large intensity variations in different image regions [30]. The fractal descriptor in wavelet domain provides sufficiently large chunk of information that help to deal with complex structures like skin lesions. In this study, the texture features have been extracted from the four color channels of red, green, blue, and luminance using wavelet-based fractal texture analysis, where

$$\text{luminance} = (0.3 \times R) + (0.59 \times G) + (0.11 \times B)$$

the input image has been decomposed into different small and large coefficient values using three-level WPD technique. The three-level WPD decomposes the input image into overall 84 nodes (4 nodes in first level, 16 nodes in second level and 64 nodes in third level). From each of the nodes of the WPD along with the original image, some statistical features such as skewness, mean intensity, entropy, minimum intensity, maximum intensity, standard deviation, variance, and also the fractal descriptor have been extracted.

Instead of using the single-valued fractal dimension, all the values in $\log(n(\delta))$ with reference to Eq. (3.26), have been considered to construct a fractal descriptor vector. To obtain the fractal descriptor from each of the node of WPD, all the nonzero coefficient values have been sorted in descending order according to the power law as shown in Eq. (3.26). The resultant fractal descriptor vector has been obtained as:

$$D_f\{|\log(D(1))|, |\log(D(2))|, |\log(D(4))|, \ldots, |\log(D(\log_2 m))|\} \tag{3.27}$$

The combination of WPD by employing classical Shannon entropy, and fractal descriptor characterizes the textural pattern of dermoscopic images by exploiting the complexity of the pixel distribution across the skin lesion area.

3.4.3.3 Results and discussions

In this study, selection of suitable mother wavelet for WPD is a critical criterion. Here, three of the most popular wavelets, namely Daubechies, Symlets, and Coiflets, have been selected as mother wavelet function. For the WPFTA, filter with shorter length such as Daubechies 1 to Daubechies 10, Symlet 2 to Symlet 8 and Coiflet 1 to Coiflet 5 have been considered. So, for the entire analysis, 22 different wavelet filters have been chosen and for each of the mother wavelet, the texture features have been extracted. The original lesion image along with its low frequency and high-frequency components obtained using WPD and fractal descriptor have been shown in Fig. 3.6. In Fig. 3.6B and C the reconstructed images from approximate and detail coefficients obtained using wavelet decomposition with 3rd order Daubechies mother wavelet,

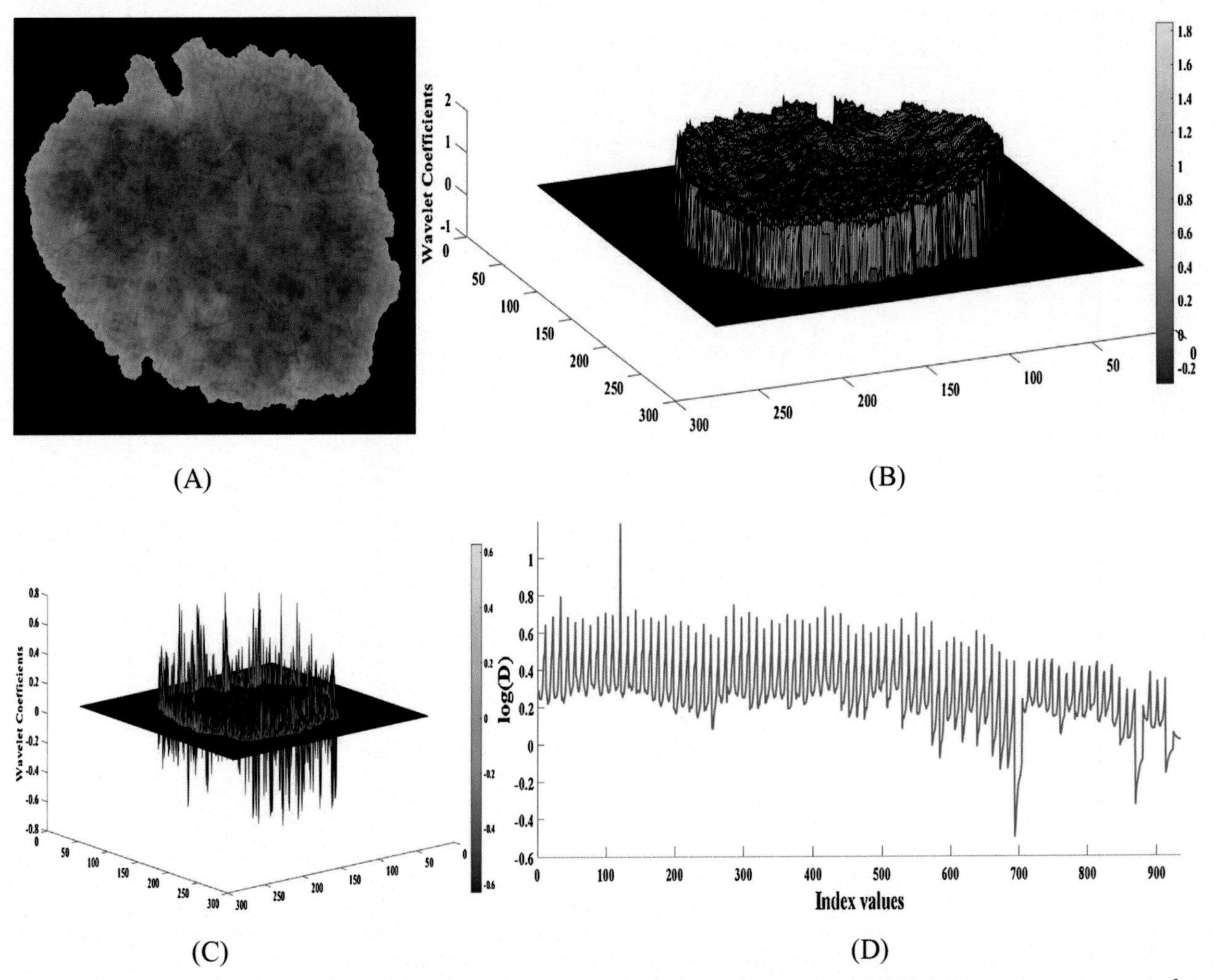

Figure 3.6 (A) Original image masked with the segmented image, (B) low frequency component, (C) high-frequency component after wavelet packet decomposition and (D) wavelet–fractal descriptor of the R channel of the original color image for 3rd-order Daubechies mother wavelet.

clearly exhibit the textural variations across the lesion area in spatial—frequency space. The reconstructed images also expose the intensity variations on the lesion area at different frequency regions which also have a self-similar nature. The high-frequency component depicts the intensity variation along the border region. Therefore the detailed coefficients yield the border features and morphological irregularity of the lesion area. However, from the approximate coefficients, the textural variations along the lesion area have been determined. Further decomposition of the approximate coefficient in the subsequent stages provides more detailed textural variations along the lesion area. The fractal descriptor as shown in Fig. 3.6D consists of all the fractality obtained from each of the 85 nodes of wavelet packet tree, including the original image. The complexity present in the lesion area at different frequency regions have been quantified using fractal descriptor.

3.4.4 Feature extraction from cross-correlation in space and frequency domain

3.4.4.1 Materials

To carry out the propounded study, only the dermoscopic images of various skin abnormalities corroborated by biopsy have been collected from different freely available databases on the internet [19—22]. The entire dataset has the dermoscopic images acquired in different imaging modalities, including both easy and difficult cases. This variability of the constituents of the dataset has helped to validate the robustness of the proposed methodology. The entire available dataset consisting of different classes of diseases, the dermoscopic images annotated as malignant or benign and categorized as melanoma, nevus, BCC, and SK. The entire dataset consists of 2879 images of melanoma, 3013 of nevus, 796 of BCC, and 513 of SK dermoscopic lesions. From the entire dataset, an image subset considering all the lesion classes has been created for the selection of kernel or informatics image patches. Considering the feature variations, overall 307 number of dermoscopic images consisting of all the four disease classes have been selected for the kernel generation. After selection of the images for kernel generation, the remaining dermoscopic images have been considered for further training and testing purpose to ensure nonoverlapping between two image sets. Fig. 3.7B has depicted the details of sample selection for kernel generation, training, and testing purposes for each class.

3.4.4.2 Kernel selection

In this present study, a set of kernels has been generated for further analysis of the skin lesions. For generating the kernels, a group of dermoscopic images from different image classes has been prepared depending on the dermatological features, in consultation with an expert dermatologist.

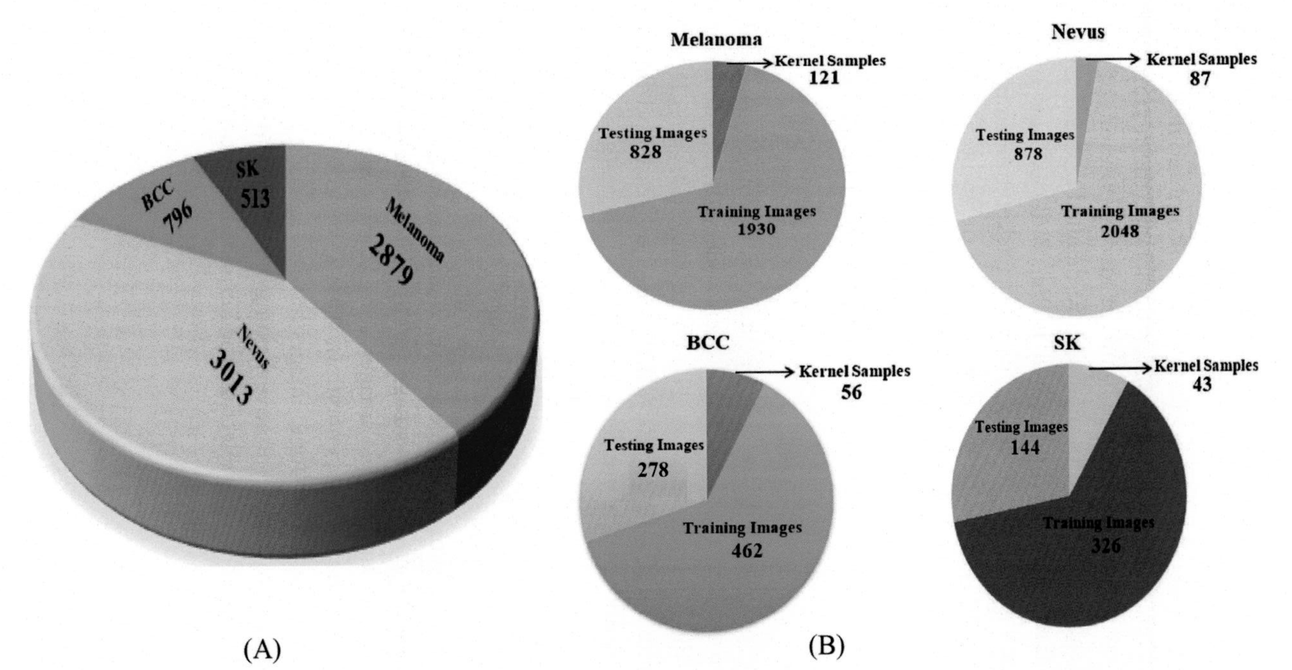

Figure 3.7 The elaborated dataset (A) original dataset, (B) dataset prepared for kernel image generation and training/testing samples of each disease classes. *BCC*, basal cell carcinoma; *SK*, seborrheic keratosis.

From each of those selected images, some specific informative regions have been extracted manually by the expert physician. These kernels contain the properties that experts use for the identification of the disease by visual inspection. For the malignant melanoma lesions, dermatological findings reveal the irregular pigment network, dots or globules of varying colors, presence of red or dark brown color and streaks, and so on. In the dermoscopic images, the identification of pseudo-pigment network and blue colored structureless area confirm the benign nevus to the dermatological experts. Similarly, the skin lesions having brain like appearance or comedo like opening ensure the categorization of seborrheic keratoses lesions. The presence of arborizing vessels and dotted blood vessels are important dermoscopic features of BCC [31]. The significant regions of skin lesions (as marked by the dermatologists) of all the four disease categories along with the explanation of corresponding dermatological features have been shown for some typical dermoscopic images in Fig. 3.8. To incorporate the scale and shift-invariant properties, the kernel images have been converted to different color spaces as discussed in Ref. [32]. Keeping in mind the invariance properties of color descriptors, the kernels have been converted to HSV, (hue-saturation-value) Opponent, C-invariant, and normalized RGB color spaces. As the intensity channel is the combination of R, G, and B channels, the feature descriptor is not invariant to light color changes. The hue color model is scale-invariant and shift-invariant with respect to the light intensity. The opponent color space has been derived from the RGB color model as follows:

$$\begin{pmatrix} O_1 \\ O_2 \\ O_3 \end{pmatrix} = \begin{pmatrix} \dfrac{R-G}{\sqrt{2}} \\ \dfrac{R+G-2B}{\sqrt{6}} \\ \dfrac{R+G+B}{\sqrt{3}} \end{pmatrix} \tag{3.28}$$

In the opponent color model the intensity information has been retained by the O_3 channel and the color information by the O_1 and O_2 channels, which are shift-invariant with respect to the light intensity. The O_1 and O_2 components of the opponent color model contain some intensity information.

The elimination of this remaining intensity information from these channels introduced the scale-invariant property as proposed by Geusebroek et al. [18]. The C-invariance [33] has been introduced by normalizing the opponent color space as (O_1/O_3) and (O_2/O_3). The illumination information has been eliminated from the color intensity channels by dividing them with intensity, making the color channels scale-invariant with respect to the light intensity. To obtain the color descriptor, the r and g chromaticity components of the normalized RGB space have been considered

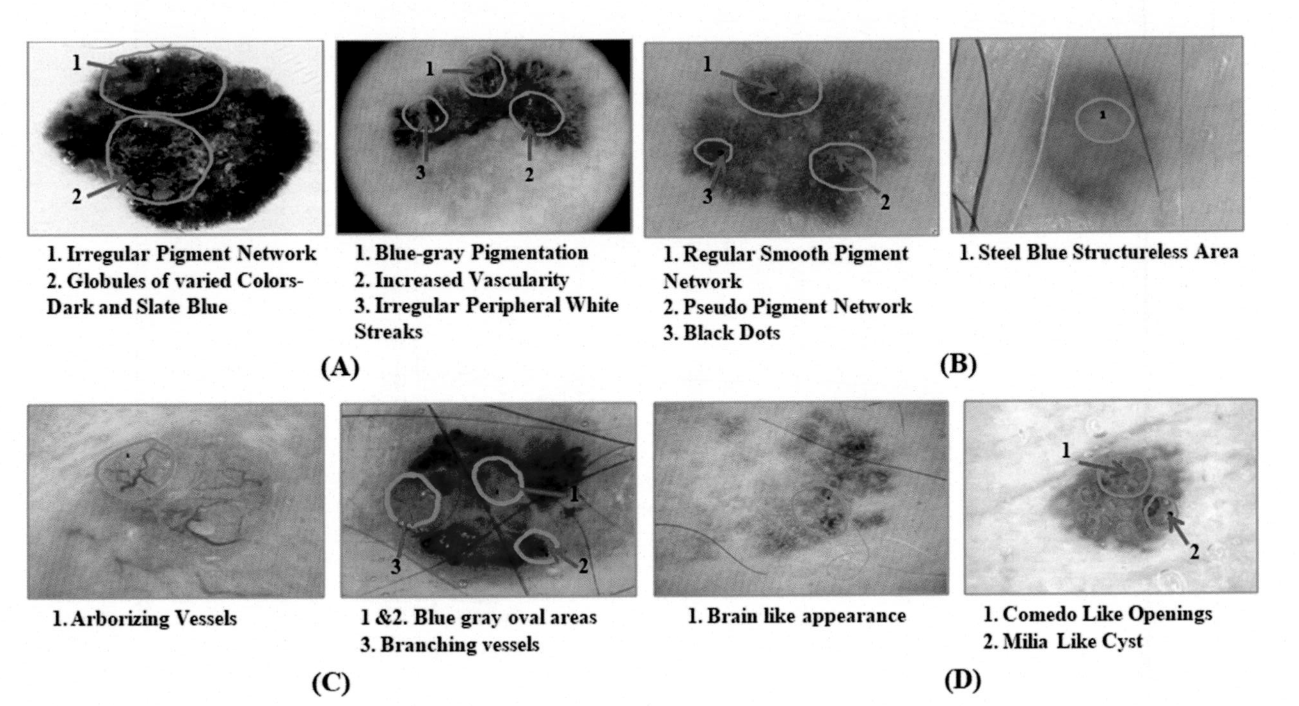

Figure 3.8 Typical examples of dermoscopic findings for the skin lesions of (A) malignant melanoma; (B) nevus; (C) basal cell carcinoma and (B) seborrheic keratosis.

due to its scale-invariant property. The R_N, G_N and B_N channels of normalized RGB color model have been computed as follows:

$$\begin{pmatrix} R_N \\ G_N \\ B_N \end{pmatrix} = \begin{pmatrix} \dfrac{R-\mu_R}{\sigma_R} \\ \dfrac{G-\mu_G}{\sigma_G} \\ \dfrac{B-\mu_B}{\sigma_B} \end{pmatrix} \tag{3.29}$$

where μ_R, μ_G, and μ_B are the means and σ_R, σ_G, and σ_B are the standard deviations of the distribution in the corresponding intensity channels of the selected kernel regions.

3.4.4.3 Methodology

The methodology of the proposed scheme is depicted in the form of a block diagram in Fig. 3.9. A bank of kernels has been prepared from the set of sample images as selected by the dermatological expert. Uniform size of the kernels has been attained by manual extraction of significant regions by the expert and subsequent conversion to different color spaces invariant to light intensity changes. The invariance properties help to incorporate skin lesion images acquired by different imaging modalities for further analysis and feature extraction. This bank of kernels contains the regional color and local morphological feature descriptors of different skin lesion classes. The set of feature descriptors have been used to obtain the informative and correlated feature regions of

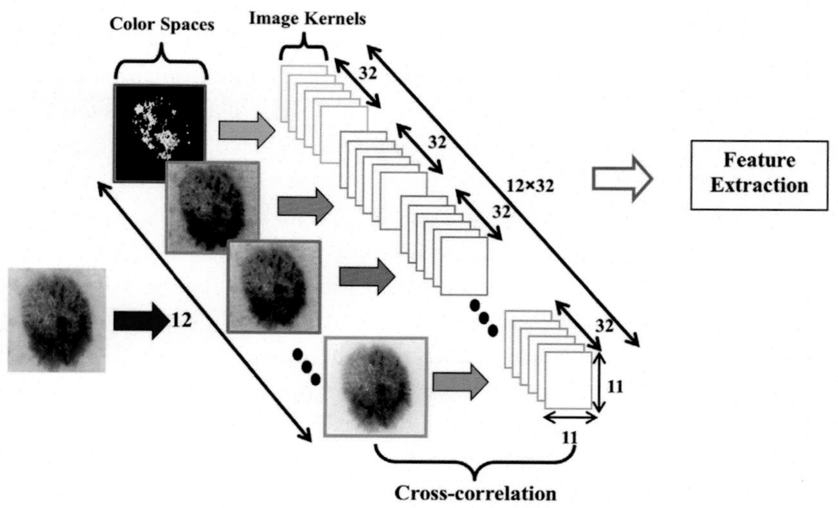

Figure 3.9 Block diagram representation of the proposed scheme.

the input sample images. After extracting the correlated feature regions from the sample images, the regional and spectral feature extraction algorithms have been deployed.

- *Region detection using cross-correlation*: For the extraction of regional feature descriptors from an image, the input RGB images have been converted to hue, opponent, C-invariant, and normalized color channels as described in previous section. In this stage, each of the color channels of the input image has been cross-correlated with the kernel images of corresponding color channels. A block of kernels represents a particular color channel, containing all the selected kernels of all lesion classes considered here. On the whole, 12 such kernel blocks have been constructed corresponding to each color channel, considering all the selected kernels of skin lesions. So, each of the input images has been converted to 12 intensity and color planes at the input stage, and each such image plane has been cross-correlated with 32 kernels in a kernel block. The selected kernels contain different significant feature regions of skin lesions. In the cross-correlation operation of the input image with the selected kernels, the similar feature regions have been identified by the higher correlation coefficient values in the region of interest. The higher correlation coefficient values imply the presence of significant features in the input image having commonalities with the selected kernel. As a consequence, if the input image has the salient feature regions akin to a particular disease class, then higher correlation coefficient values will be assigned to that region. For the extraction of the significant feature regions, the correlation coefficient values have been selected in the range of 0.7–1.
- *Feature extraction*: The cross-correlation of the input dermoscopic image with the selected kernels has identified significant regions of the skin lesion area having higher similarity indices with the corresponding kernels. So, the resultant cross-correlated image contains smaller subregions having similar dermoscopic features of the same disease class. The regional feature extraction has guided towards the extraction of similar color information as the kernels and the frequency domain analysis has provided more detailed spectral information. From each of the extracted meaningful image segments, regional spatial, and spectral features have been extracted using statistical feature extraction and cross-spectrum analysis technique, respectively.
- *Spatial feature extraction*: Referring to Fig. 3.9, it can be seen that each input image has been cross-correlated with 32 kernels of different dermoscopic features in each of the color plane. So, the number of extracted meaningful regions varies for different samples of skin lesion images based on the class commonality of the input image and the kernel. The spatial information of each of the subregions describes the color feature of the entire skin lesion area. However, the nonuniformity of the extracted subregions of each image has introduced difficulty in the formation of uniform feature vector for the classification of skin abnormalities. Such a complicated issue has been addressed here by introducing a spatial feature extraction technique. The histogram depicts the frequency of occurrence of intensity in an image. The histogram corresponding to each of the extracted subsegments of the lesion describes the intensity

distribution of that region. For every color channel, in the histogram of each subregion, the regional intensity value having maximum frequency of occurrence has been considered to be the dominating intensity information of that region. Considering the prevailing intensity information as the key feature of the corresponding subregions, a regional key intensity feature descriptor has been constructed. The key feature descriptor has illustrated the significant intensity distribution along the region of interest of the dermoscopic image in the corresponding color channel under consideration. The extraction of statistical features like maximum, minimum, mean and standard deviation from key feature descriptors of the corresponding color channels have been considered as the representative features of that dermoscopic image.

- *Cross-spectrum-based feature extraction*: The frequency domain analysis has been extensively used for the extraction of more detailed information from an image. Fourier transform depicts the geometric characteristics of a spatial domain image by all the frequencies present in that image. The 2D discrete Fourier transform of an image I of size $P \times Q$ is expressed as follows:

$$F(u,v) = \frac{1}{PQ}\sum_{p=0}^{P-1}\sum_{q=0}^{Q-1} I(p,q)\exp\left[-2\pi j\left(\frac{pu}{P} + \frac{qv}{Q}\right)\right] \quad (3.30)$$

where $u = 0, 1, \ldots, P - 1$ and $v = 0, 1, \ldots, Q - 1$ Here, the cross-correlation operations of the input sample image with the selected kernels have demonstrated the degree of similarity of the feature descriptors with the correlation coefficient values. The Fourier transformation of the 2D cross-correlation describes the similarity indices in the frequency domain. So the cross-spectrum analysis determines the spectral similarity between the input image and the selected kernels.

The cross-spectrum can be expressed as:

$$S_C = \sum_{u=0}^{S-1}\sum_{v=0}^{T-1} e^{-2\pi ju/S} e^{-2\pi jv/T} I_C \quad (3.31)$$

where I_c is the resultant cross-correlated image of size $S \times T$ as given in Eq. (3.30) and j is the imaginary unit. The magnitude and phase information have been captured from the elements of the resultant cross-spectrum S_C. The magnitude of the Fourier transform contains most of textural information of the spatial domain image. The amplitude spectrum has been expressed as,

$$\alpha_{I,K}(s,t) = (\sigma_{I,K}(s,t)^2 + \varphi_{I,K}(s,t)^2)^{1/2} \quad (3.32)$$

where $S_C = \sigma_{I,K}(s,t) + j\varphi_{I,K}(s,t)$ is the resultant cross-spectrum, while $\sigma_{I,K}$ and $\varphi_{I,K}$ are its real and imaginary parts, respectively. The magnitude of the resultant cross-spectrum of the input image has been analyzed to obtain the common attributes of the input dermoscopic image and the selected kernel patches. The degree of spectral similarity between the input image and the selected kernels have manifested themselves in the magnitude spectra, as higher magnitude values. The elevated magnitude values represent the most significant commonality of the feature descriptor present in the smaller subsegment of the lesion area of that specific frequency region. The distribution of the cross magnitude spectrum coefficients has revealed the similarity between the spectral features of the input dermoscopic image and the kernels. The higher magnitude values near the center regions of the magnitude spectrum delineate the spectral similarity of the input image and the selected kernel in the low frequency region. Similarly, the magnitude values located away from the center, describe the resemblance in the high-frequency regions. Considering the even function symmetry of the Fourier transformation, the magnitude spectra along the horizontal direction have been considered as the befitting representative of the spectral feature of the region of interest. The entire feature vector has been constructed by considering the statistical spatial features extracted from significant image regions and the spectral features as the magnitude spectra along the horizontal direction. The feature vector contains the region-based spatial and spectral features, which has demonstrated the distribution of significant features along the skin lesion area.

3.4.4.4 Results and discussion

In Fig. 3.10 the 2D cross-correlograms portray the correlation coefficient values of the input image with the selected kernels. Here, the substantially peaky regions describe the higher regional similarity of the input image with a specific kernel patch. Higher correlation coefficient values get assigned when similarity between skin lesion area and the kernel patches get increased. The pixels having the correlation coefficients within the range of 0.7−1 have been considered to constitute that part of the region of interest having higher similarity index with a kernel of a particular disease class. Thus the cross-correlation of the input image with the set of kernels of different disease classes has extracted the meaningful regions from each of the image for further feature extraction automatically, depending on the similarity indices. In Fig. 3.11, some of the regions having higher similarity indices with the selected kernel feature descriptors have been marked with squares. The figures indicate that the highly correlated regions are concentrated in the skin lesion area. Thus the assignment of the higher correlation coefficients in the regions of interest, significantly describes the presence of similar feature descriptor of the corresponding skin disease, represented by the kernel patches. Fig. 3.12 depicts some examples of even function symmetric cross-spectra of the input dermoscopic images with the kernels. These examples are representative cases, where the cross-spectra indicate the similar spectral features of the input image and the kernel at low frequency, high frequency, and at both low- and high-frequency regions, respectively. This frequency domain analysis has helped to obtain feature descriptors providing more detailed insight of the smaller subsegments of the lesion area.

Figure 3.10 Typical examples of 2D cross-correlograms obtained from cross-correlation operations of the input image with the selected kernel patches.

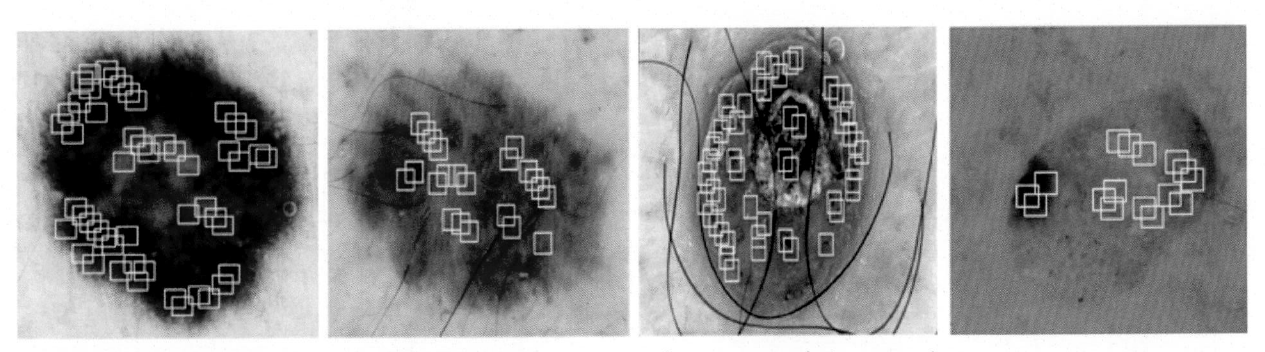

Figure 3.11 Examples of some detected regions of the skin lesion area after cross-correlation operation.

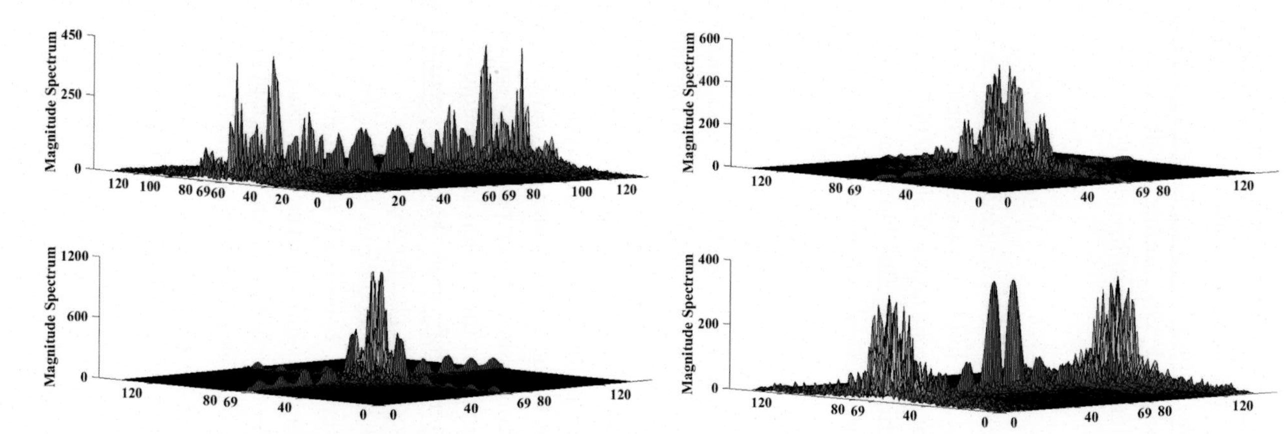

Figure 3.12 Some typical examples of magnitude spectrum of dermoscopic images.

3.4.5 Sparse autoencoder-based feature extraction

3.4.5.1 Materials

In this study, four different types of diseases have been considered as melanoma, nevus, BCC, and SK. For this study, the histopathologically confirmed 2172 number of dermoscopic images of malignant melanoma, 4442 numbers of benign melanocytic nevus, 419 numbers of SK lesion images, and 596 number of dermoscopic images of BCC have been considered from the International Skin Imaging Collaboration challenge dataset [22].

3.4.5.2 Methodology

Neural network with multiple hidden layers can be considered to solve the classification problems from image dataset. The hidden layers of the neural network are responsible for providing enriched features at different levels of abstraction. Here, sparse autoencoder has been used as a feature extracting tool for the skin lesion images. To acquire more important and distinguishable features, several hidden layers are stacked in cascaded form. From the figure, it can be seen that the autoencoder consists of input layer $(x_1, x_2, \ldots, x_K)$, hidden layer and output layer $(\hat{x}_1, \hat{x}_2, \ldots, \hat{x}_K)$. The encoder is associated with the mapping of input information to the hidden representation. On the other hand, decoder is used to reconstruct the input information from the hidden representation. The hidden layer of the first autoencoder $(h_1^{(1)}, h_2^{(1)}, \ldots, h_L^{(1)})$ is considered as the input to the second autoencoder with a hidden layer $(h_1^{(2)}, h_2^{(2)}, \ldots, h_P^{(2)})$ having same or reduced number of nodes. A schematic representation of a stacked auto-encoder is shown in Fig. 3.13. This arrangement of using the feature descriptor or the hidden layer of one autoencoder as an input to the next subsequent stage of the autoencoder has constructed a stacked autoencoder. The hidden layer of the second autoencoder provides a set of representative feature in reduced form. Number of nodes in the hidden layer is equivalent to the number of features extracted from the input information. However, training a neural network with multiple hidden layers is very much challenging in practice. Therefore to train a multilayer neural network or stacked autoencoder model, one layer is trained at a time.

Here, the dermoscopic images have been considered as inputs to the autoencoder module having hundred neurons as hidden layer. Each pixel of the input image has been connected with each neuron of the hidden layer. A dropout layer is incorporated to inactive the input nodes having zero value, so that those nodes should not be used in forward propagation. Training the first autoencoder with the input data, a learned feature vector is obtained as the hidden layer. Such hidden layers have been used as input to the next autoencoder with 50 numbers of neurons. After training the first autoencoder, in similar way the training of second stage of autoencoder provides a reduced learned feature set. This feature set is considered to be the input in the construction of last and final stage of the stacked autoencoder having 10 nodes in the hidden layer. The learned weight values of the last hidden layer of the stacked autoencoder have been considered as the extracted features from that particular dermoscopic image. The reduced number of nodes in the

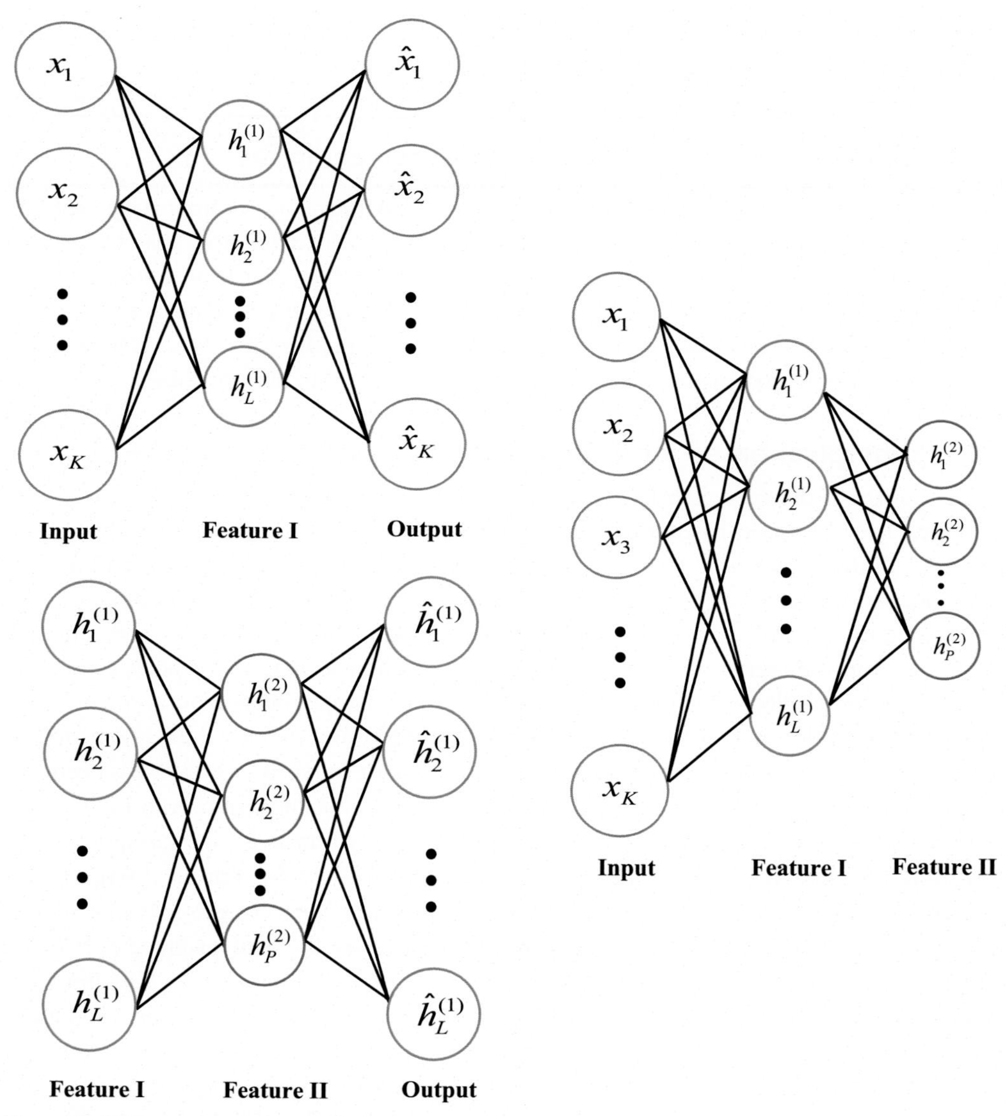

Figure 3.13 Structure of the stacked auto-encoder model.

subsequent stages of the stacked autoencoder ensures more smaller representation of the input images with significant information. The L2 regularization parameter has been selected as 0.002 to regularize the weight values for reducing the likelihood of model overfitting. To control the sparsity of the output from the hidden layer, the parameter for sparsity regularizer is set to a very small value of 0.15. The sparsity regularizer attempts to enforce a constraint on the sparsity of the output from the hidden layer. Employing the stacked autoencoder-based technique, the representative features of the input image have been extracted and selected subsequently. This feature set can be further considered to feed to a classifier model for the classification of the diseases.

3.5 Application of color feature extraction techniques

The extraction of visual characteristics from a medical image is an important step for computer-aided diagnosis. There are two fundamental visual descriptors that have been explored for medical image processing: texture and color. The extraction of color features from dermoscopic images is challenging due to the various imaging modalities and because different skin lesions have their own typical set of colors. Presence of various color regions in the dermoscopic images are considered for differentiating skin abnormalities.

3.5.1 Superpixel-based local color feature extraction technique

In the literature, different global feature extraction techniques have been explored for the color features extraction-based on statistical measures from color histogram, which have inherent advantages for scale and rotation invariant features extraction. Making use of these advantages, in this study the regional color features have been extracted by grouping the pixels in some meaningful superpixel regions [34]. As the skin lesion area contains different colors throughout the region, the local color feature extraction technique has been considered to obtain more detailed color feature descriptor. In the literature, different authors have considered various color models for the extraction of color features. Among these color models, the HSI (hue—saturation—intensity) model describes a color in terms of how it has been discerned by the human eye. Naturally, more effective pure color information can be obtained in terms of hue and saturation than in terms of addition of subtractive color components. In this study, the HSI color model has been used to obtain the color feature descriptor. To construct the detailed color feature descriptor, the skin lesion area has been split into smaller subregions by replacing pixels with superpixels. To generate the superpixels, simple linear iterative clustering (SLIC) method [34] has been employed. Using SLIC algorithm, an approximately equally sized, desired number of superpixels (n) have been generated from the collection of equally informative and meaningful pixels of an image (of M number of pixels). The desired number of superpixels has been considered as the only parameter of this algorithm. The superpixel generation has been initiated with initializing the n number of seed locations in the HSI color model as: $SP_n = [H_n, S_n, I_n, x_n, y_n]^T$, with a sample size of K, where $K = \sqrt{M/S}$. The seed locations have been identified as the lowest gradient position in 3×3 neighbors. In this study, the SLIC algorithm has been implemented to construct the superpixels in the hsixy color image plane space. The distance D has been computed between a pixel i and seed within the search location $2K \times 2K$ as follows:

$$D_{sp} = \sqrt{(H_i - H_n)^2 + (S_i - S_n)^2 + (I_i - I_n)^2}$$

$$D_l = \sqrt{(x_i - x_n)^2 + (y_i - y_n)^2}$$

$$D = \sqrt{D_{sp}^2 + \frac{D_l^2}{K}\delta^2}$$

In Eqs. (3.31) and (3.32), D_{sp} and D_l represent the color proximity and spatial distances, respectively. The combination of two distances have been used to calculate the distance D as in Eq. (3.21), where the spatial proximity has been normalized by the maximum distance within a superpixel (K) and the multiplying factor δ has been introduced to weigh the relative importance between the color similarity and spatial proximity.

The SLIC algorithm generates the desired number of superpixels depending on the color information present in the skin lesion area. To ensure that the generated superpixels are well confined in the boundary of the region of interest, the original image (in HSI color space), masked with the segmented image has been considered as an input to the SLIC algorithm. Examples of dermoscopic images with their HSI color model and superpixel regions have been given in Fig. 3.14. To extract the regional color features, the size of the superpixels has been made smaller and compact by giving more relative importance to color proximity than the spatial distances. The effect of

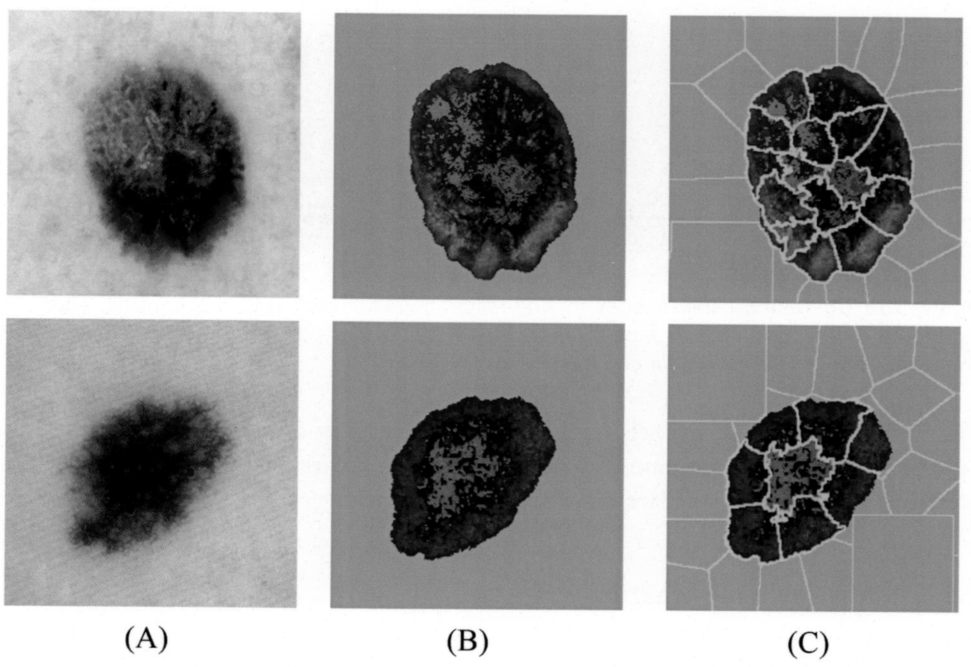

Figure 3.14 Dermoscopic images (A) in RGB color plane; (B) in hue–saturation–intensity color model and (C) superpixel regions.

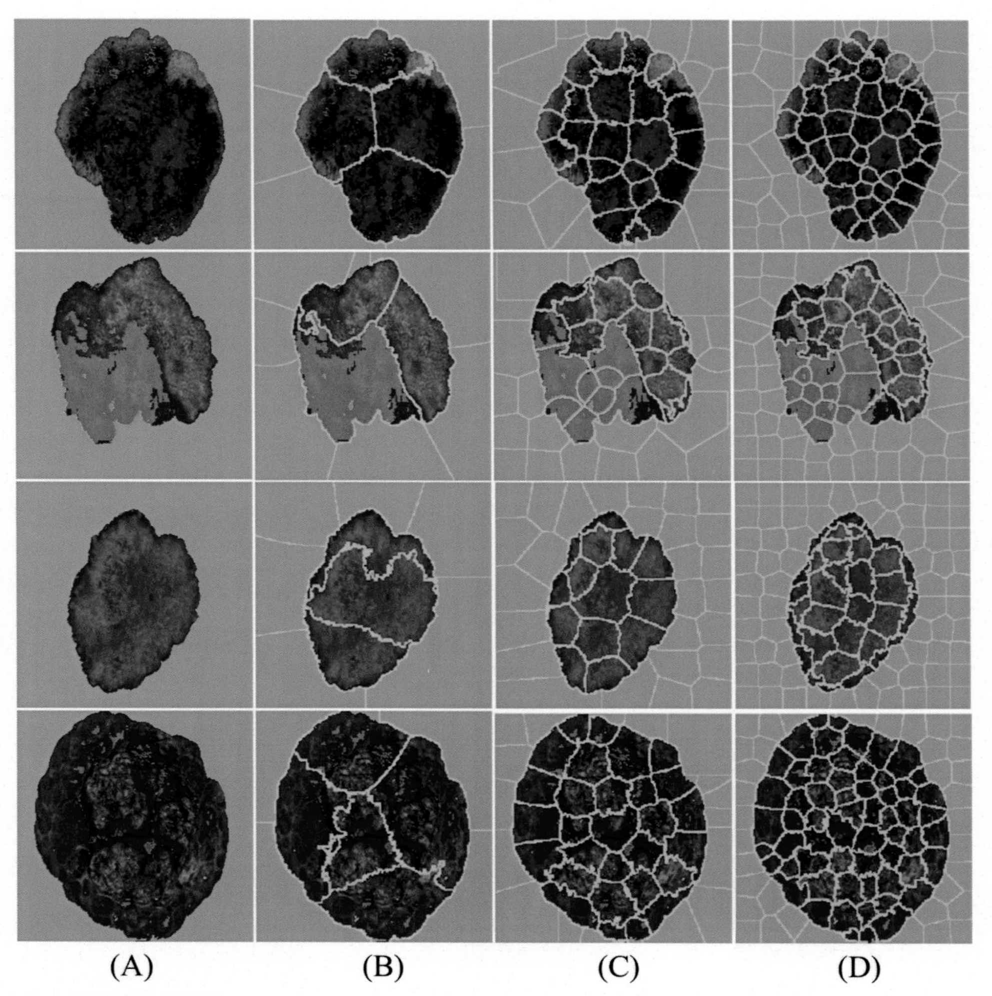

Figure 3.15 (A) Dermoscopic images in hue–saturation–intensity (HIS) color plane; superpixel regions for different number of superpixels as (B) 10, (C) 50, (D) 100 (upper two rows are for melanoma and lower two rows are for common nevus images).

different number of superpixels has been demonstrated in Fig. 3.15. From each of the superpixel, the color histogram features, namely mean, variance, maximum, minimum, skewness, entropy and standard deviation, are extracted.

3.5.2 Statistical color feature extraction

For the extraction of color features, some statistical measures have been considered. These are minimum, maximum, standard deviation, skewness, entropy, mean, and variance of nine color channels—R, G, B, H, S, V, L, A, and B from each sample (symbols carry the usual meaning for a color image). So, in all, 63 color features have been extracted from each sample.

3.6 Conclusion

In this chapter, digital signal processing techniques are used to develop feature extracting tools for the quantitative analysis of dermoscopic images. For the identification of various skin abnormalities, morphological, textural, and color features are extracted by employing signal processing tools. Statistical morphological features are utilized to estimate the area of the affected region and further spreading of the diseases. Abnormal structural properties are quantified by determining the border irregularity measuring the fractal dimension and wavelet-based fractal analysis. Qualitative textural analysis of the skin lesion is not sufficient to describe the textural complexity. For the texture feature extraction, FRTA algorithm is introduced to determine the textural complexity by extracting regional texture information. WPFTA is reported to construct wavelet–fractal descriptor for the quantification of textural irregularity of the lesion area. Cross-correlation-based spatial and spectral feature extraction technique is employed for the identification of similarity-based texture feature descriptor. Unsupervised learning-based sparse autoencoder is used to obtain reduced set of feature descriptor from the original dermoscopic images. Identification of significant color regions and corresponding feature extraction lead to the appropriate classification of skin diseases. Superpixel-based local color feature extraction technique is reported to quantify the color features from skin lesion area. Quantification of significant information can be considered by the expert dermatologist for the identification and further monitoring of the disease. Appropriate feature extraction is essential for the development of classification models for further classification of the disease.

References

[1] C. Barata, M. Ruela, M. Francisco, T. Mendonc, J.S. Marques, Two systems for the detection of melanomas in dermoscopic images using texture and color features, IEEE Syst. J. 8 (3) (2014) 965–979.

[2] M. Rastgoo, R. Garcia, O. Morel, F. Marzani, " Automatic differentiation of melanoma from dysplastic nevi, Comput. Med. Imaging Graph. 43 (2015) 44–52.

[3] R. Garnavi, M. Aldeen, J. Bailey, Computer-aided diagnosis of melanoma using border and wavelet-based texture analysis, IEEE Trans. Inf. Tech. BioMed. 16 (6) (2012) 1239–1252.

[4] R. Kasmi, K. Mokrani, Classification of malignant melanoma and benign skin lesions: implementation of automatic ABCD rule, IET Image Process 10 (6) (2016) 448–455.

[5] J. Kawahara, G. Hamarneh, Fully convolutional neural networks to detect clinical dermoscopic features, IEEE J. Biomed. Health Inform 23 (2) (2019) 578–585.

[6] I. González-Díaz, DermaKNet: incorporating the knowledge of dermatologists to convolutional neural networks for skin lesion diagnosis, IEEE J. Biomed. Health Inform 23 (2) (2019) 547–559 (2018).

[7] B.B. Mandelbrot, Fractals: Form, Chance and Dimensions, Freeman, San Francisco, CA, 1977.
[8] K. Falconer, Fractal Geometry Mathematical Foundations and Applications, second ed., Wiley, 2003.
[9] F. Hausdorff, Dimension and äußeres Maß, Math. Annal 79 (1–2) (1918) 157–179.
[10] T. Higuchi, Approach to an irregular time series on the basis of the fractal theory, Physics D 31 (1988) 277–283.
[11] M. Katz, Fractals and the analysis of waveforms, Comput. Biol. Med. 18 (3) (1988) 145–156.
[12] R.C. Gonzalez, R.E. Woods, Digital Image Processing, third ed., Pearson, 2014.
[13] J. Gilles, Empirical wavelet transform, IEEE Trans. Signal Proc. 61 (2013) 16.
[14] J. Gilles, G. Tran, S. Osher, 2D Empirical transforms. Wavelets, ridgelets, and curvelets revisited, SIAM J. Imgaging Sci 7 (1) (2014) 157–186.
[15] W. Sun, S. Shao, R. Zhao, R. Yan, X. Zhang, X. Chen, A sparse auto-encoder based deep neural network approach for induction motor fault classification, Measurement 89 (2016) 171–178.
[16] J.P. Lewis, Fast Normalized Cross-Correlation, Industrial Light Magic, 2019.
[17] S. Chandaka, A. Chatterjee, S. Munshi, Cross-correlation aided support vector machine classifier for classification of EEG signals, Expert Syst. Appl. 36 (2009) 1329–1336.
[18] J.M. Geusebroek, R. van den Boomgaard, A.W.M. Smeulders, H. Geerts, Color invariance, IEEE Trans. Pattern Anal. Mach. Intell. 23 (12) (2001) 1338–1350.
[19] International Dermoscopy Society. http://www.dermoscopy-ids.org.
[20] Dermoscopy Atlas. http://www.deroscopyatlas.com.
[21] T. Mendonça, P.M. Ferreira, J. Marques, A.R.S. Marcal, J. Rozeira, PH2—a dermoscopic image database for research and benchmarking, in: 35th International Conference of the IEEE Engineering in Medicine and Biology Society, July 3–7, 2013, Osaka, Japan.
[22] D. Gutman et al., Skin lesion analysis toward melanoma detection: Achallenge at the international symposium on biomedical imaging (ISBI) 2016, hosted by the international skin imaging collaboration (ISIC), 2016 [Online] Available: <https://arxiv.org/abs/1605.01397>.
[23] A.F. Costa, G. Humpire-Mamani, A.J.M. Traina, An efficient algorithm for fractal analysis of textures, in: Proceedings of the XXV SIBGRAPI Conference on Graphics, Patterns and Images, pp. 39–46, 2012.
[24] N. Otsu, A threshold selection method from gray-level histograms, IEEE Trans. Syst. Man Cybern. 9 (1979).
[25] I. Guyon, S. Gunn, M. Nikravesh, L.A. Zadeh, Feature Extraction Foundations and Applications, vol. 207, Springer, 2006.
[26] B.B. Chaudhuriand, N. Sarkar, An efficient approach to compute fractal dimension in texture image, Pattern Recognit 1 (1992) 358–361.
[27] N. Sarkarand, B.B. Chaudhuri, An efficient differential box-counting approach to compute fractal dimension of an image, IEEE Trans. Syst. Man Cyber 24 (1) (1994) 115–120.
[28] S. Deppaand, T. Tessamma, Fractal features based on differential box counting method for the categoritazion of digital mammograms, Int. J. Comput. Inform. Syst. Ind. Manage. Appl. 2 (2010) 011–019.
[29] C.L. Jones, H.F. Jelinek, Wavelet packet fractal analysis of neuronal morphology, Methods 24 (4) (2001) 347–358.
[30] J.B. Florindo, O.M. Bruno, Texture analysis by fractal descriptors over the wavelet domain using a best basis decomposition, Physics A 444 (2016) 415–427.
[31] P. Kharazmi, Md. I. AlJasser, H. Lui, Z.J. Wang, T.K. Lee, Automated detection and segmentation of vascular structures of skin lesions seen in dermoscopy, with an application to basal cell carcinoma classification, IEEE J. Biomed. Health Inform. (2017). Available from: https://doi.org/10.1109/JBHI.2016.2637342.
[32] K.E.A. van de Sande, Evaluating color descriptors for object and scene recognition, IEEE Trans. Pattern Anal. Mech. Intell. 32 (9) (2010) 1582–1596.
[33] G.J. Burghouts, J.M. Geusebroek, Performance evaluation of local color invariants, Comput. Vis. Image Underst. 113 (2009) 48–62.
[34] R. Achanta, A. Shaji, K. Smith, A. Lucchi, P. Fua, S. Süsstrunk, SLIC superpixels compared to state-of-the-art superpixel methods, IEEE Trans. Pattern Anal. Mech. Intell. 34 (11) (2012) 2274–2281.

CHAPTER 4

Feature selection and classification

4.1 Introduction

Among all types of skin cancers, malignant melanoma is the deadliest. Despite worldwide efforts to standardize the dermatoscopic findings so that such findings are not subjective and they become reproducible, there exists a scope for computer-aided diagnostic system to assist the clinicians and dermatologists in decision making, especially in the cases of investigating an outsized number of patients' data in a shorter duration of time. Similarity between different skin lesions and complex appearance in terms of shape, texture, and color make it difficult to differentiate the skin diseases. Similarity in the manifestations of melanoma and dysplastic nevi, which are both characterized by melanocytic skin lesion, makes the early diagnosis not an easy task for the dermatologists and other physicians. In some cases, dysplastic nevi may develop into melanoma, whereas some patients with atypical mole may be treated for melanoma due to their close dermatological similarities. Basal cell carcinoma (BCC) is another common type of malignant epidermal lesion [1].

Maglogiannis and Delibasis [2] have extensively described different dermatological feature quantification methodologies, related to shape, texture, and color of the skin lesion in a computer-aided diagnostic system for the skin disease identification. Shimizu et al. [3] have reported a layered model for classification of four different types of skin diseases, namely melanoma, nevi, BCC and seborrheic keratosis (SK) based on the lesion color, subregion, and textural features. Using the layered model, the classification performance has been achieved with the detection rate of 90.48%, 82.51%, 82.61% and 80.61% for melanoma, nevi, BCC, and SK, respectively. Rastgoo et al. [4] have proposed a classification framework for the differentiation melanoma from dysplastic nevi, considering the local and global shape, texture, and color information from dermoscopic images, using support vector machine (SVM), gradient boosting, and random forest classifier for individual and combined features. Highest sensitivity of 98% and specificity of 70% have been achieved in the reported random forest-based classification framework for the differentiation of melanoma using texture features. Performance of two different systems based on global and local texture and color features of lesion in dermoscopic images have been compared with a performance baseline of 96% sensitivity and 80% specificity for global method and 100% sensitivity and 75% specificity for local method, for the identification of melanoma in Barata et al. [5]. Sáez et al. [6] have proposed a comparative study of supervised

Recent Trends in Computer-aided Diagnostic Systems for Skin Diseases
DOI: https://doi.org/10.1016/B978-0-323-91211-2.00001-9

classification schemes for the determination of different stages of the diseases, based on melanoma thickness from dermoscopic images and have achieved 77.6% classification accuracy using a method combining logistic regression with artificial neural network. Oliviera et al. [7] have evaluated different ensemble classification models based on formation of the feature subset from shape properties, color variation, and texture analysis. The computational system has achieved 94.3% accuracy, 91.8% sensitivity, and 96.7% specificity.

Literature reports deep learning-based techniques as efficient tool for skin disease analysis. Ravì et al. [8] have demonstrated advancement in the machine learning approach, deploying deep learning techniques in the fields of translational bioinformatics, medical imaging, pervasive sensing, medical informatics, and public health. In Ref. [9], Iván González-Díaz has proposed a computer-aided diagnostic system using convolutional neuralnetwork, incorporating the dermatologist's knowledge. In the proposed DermaKNet model, an automatic dermoscopic structure segmentation and diagnosis module has been introduced, considering melanoma, nevus and SK diseases. Harangi has classified melanoma, nevus, and SK images using ensembles of deep convolutional neural networks [10]. Kawahara and Hamarneh have proposed a fully convolutional neural network architecture, interpolating feature maps from a number of intermediate layers of the network for the detection of clinical dermoscopic features [11]. A convolutional–deconvolutional neural network has been introduced by Yuan and Lo for the segmentation of skin lesions [12]. Kawahara et al. has proposed a multitask deep convolutional neural network, trained on clinical and dermoscopic images, and patient metadata for the diagnosis of skin diseases based on melanoma seven-point check list criterion [13].

In this chapter, classification techniques for identification of various skin abnormalities have been discussed. As discussed in the previous chapter, various signal processing tools have been employed to quantify morphological, texture, and color properties of the skin lesion. For the identification of skin diseases based on that quantitative information, classification models have been developed. Prior to the classification stage, important and differentiating features have been selected by introducing efficient feature selection algorithm. In this book, the classification problems have been categorized into three sections depending on the number of disease classes. SVM classifier is widely explored to develop binary and multiclass classification technique. For the multiclass classification technique, ensemble model of classifiers have been developed. Feature selection algorithm is chosen according to the length of feature vector. To select reduced set of features from a large number of features set, a support vector machine-based recursive feature elimination (SVM-RFE) technique is used. The automatic correlation bias reduction (CBR) algorithm has been introduced to obtain reduced number of features from wide range of feature descriptor.

4.2 Machine learning: an overview

4.2.1 Learning machine: an idea

The computational intelligence system is capable of acquiring adequate knowledge from data analysis. The concept of learning machines is the fundamental element of intelligent systems encompassing different groups of efficient algorithms including machine learning, soft computing, evolutionary algorithm, and so on. Learning machines is an algorithm demonstrating a learning model that transforms objects from data or information domain (X) to target or class domain (Y) as a function:

$$f:X \rightarrow Y \tag{4.1}$$

The data domain and the set of targets are obtained from the definition of the problem under consideration. Following the definition of the problem, the ability to learn is one of the distinctive attributes of leaning machines algorithms.

From the literature [14], learning can be defined as follows: "Learning process includes the acquisition of new declarative knowledge, the development of motor and cognitive skills through instruction or practice, the organization of new knowledge into general, effective representations, and the discovery of new facts and theories through observation and experimentation." Therefore machine learning algorithms can be classified based on the underlying learning strategies employed. The primary entities of machine learning are the teacher who possess sufficient knowledge to find the suitable solution of a problem and the learner, who should learn the knowledge to perform the task. The learning strategies can be differentiated according to the amount of inference the learner draws from the available information provided by the teacher. Depending on the amount of effort required for the learner and the teacher, the learning types can be categorized in four different forms. In rote learning, the new knowledge is directly induced to the learner without any need of further inference or transformation of the knowledge. In the case of learning from instruction type, the learner acquires knowledge from the teacher as well as other sources of knowledge and integrate all the new information for effective use. Here the learner makes some inferences with significant cognitive burden to the teacher to incrementally elevate its knowledge. For the learning by analogy, the learner requires to acquire new set of knowledge by transforming and boosting the existing knowledge to deal with new situation. Therefore the learning by analogy requires more inference as rote learning and learning from instruction. Today, the concept of learning from examples technique has become so popular that this learning process implicates itself as simply learning. Therefore in other way, the learner and examples are referred to as learning machine and data, respectively. In this context, learning can be considered as processes to search the parameters of the model f using learning algorithms to obtain desired solution of the problem [15–17]. The learning algorithm learns from a data sequence D in X or $X \times Y$ domain as the following:

$$D = \{x_1, x_2, \ldots, x_s\} = X \tag{4.2}$$

$$D = \{\langle x_1, y_1 \rangle, \langle x_2, y_2 \rangle, \ldots, \langle x_s, y_s \rangle\} = \langle X, Y \rangle \tag{4.3}$$

Usually, x_i corresponds to the feature vector containing the explanatory information of the corresponding class labels y_i. The definition stated in Eq. (4.2) reflects the unsupervised learning problem (learning without teacher) based on the information (x_i) contained in multidimensional vector. On the other hand, learning algorithm based on the pairs $\langle x_i, y_i \rangle$ describes the supervised learning problem (learning from teacher). In the problem of supervised learning, the sample of input output pairs ($\langle x_i, y_i \rangle$), called the training sample (or training set), the task is to find a deterministic function that maps any input to an output that can predict future input output observations, minimizing the errors as much as possible. Supervised learning can be categorized into classification and regression learning depending on the types of outputs. In the classification learning or simply classification, the objective is to verify whether in output space the two elements of the outputs are equal or not. Here, the output space refers to the class labels. The class label $y = 2$ corresponds to the binary classification problem and $y > 2$ is multiclass classification problem. In more convenient form, the class labels are represented as $y = \{+1, -1\}$ or $y = \{0, 1\}$ or $y = \{1, \ldots, C\}$.

4.3 Support vector machine: a brief introduction

The SVM was introduced by Boser et al. and thoroughly examined by Vapnik [18,19]. The basic idea behind the SVM is to develop a separating hyperplane with disparate samples of different classes with maximal distance between the hyperplane and the closest training data points [20]. The learning algorithm learns from a data sequence, $\{x_i, y_i\}$, $i = 1, 2, \ldots, m$. In the pair $\{x_i, y_i\}$, y_i is the desired output value for the given feature vector x_i, where $y_i \in \{-1, +1\}$ and $x_i \in R^d$. For linear learning machine, an optimal hyperplane is constructed using the optimal values of weight vector w and bias b for the model f, where

$$f(x) = w^T x + b$$

Thus depending on the condition $f(x) > 0$ for an input feature vector x, the positive and negative labels are assigned to the classes. The point x that satisfies the condition $w^T x + b$ defines the hyperplane, where w is normal to the hyperplane, $|b|/||w||$ is the perpendicular distance from the hyperplane to the origin, and $||w||$ is the Euclidean norm of w. The distance between the hyperplane and the closest training data points are called support vectors. For maximization of the minimum distances between the support vectors and the hyperplane, the objective function has been considered as

$$\max_{w,b}\min\{||x - x_i||:w^T x + b = 0, \quad i = 1, \ldots, m\}$$

For every x_i:$y_i[w^T + b] \geq 1$, so that the width of the margin is equal to $2/||w||$. This goal can be restated as the optimization problem of another objective function $\tau(w)$: $\min_{w,b} \tau(w) = (1/2)w^2$. To solve it, a Lagrangian is constructed as follows:

$$L(w, b, \alpha) = \frac{1}{2}||w||^2 - \sum_{i=1}^{m} \alpha_i(y_i[x_i^T w + b] - 1)$$

where $\alpha_i > 0$ are Lagrange multipliers. Its minimization leads to $\sum_{i=1}^{m} \alpha_i y_i = 0$, that is, $w = \sum_{i=1}^{m} \alpha_i y_i x_i$. The samples corresponding to nonzero α's are known as support vectors.

4.4 Recursive feature elimination based feature selection technique

4.4.1 SVM-RFE feature selection algorithm

The feature selection method helps to find the most appropriate and informative set of features, thereby increasing the classification accuracy. It reduces the computation time by slashing the data size. Considering all the variants, the recursive feature elimination (RFE) method combined with nonlinear SVM has been found to be effective in various applications [21,22]. Starting with all features, according to a selection criterion, the backward elimination methods iteratively remove features from the data until a stopping condition is reached. RFE is a weight-based backward elimination or selection algorithm described by Guyon et al. [19]. The essence of the RFE technique is to first train the SVM classifier with the entire feature set and consequently identifies those that cause least decrement of the margin. This is indicated by a prespecified stopping condition. The hyperplane is constructed by the SVM model to differentiate different classes. The ranking criterion for feature k is the square of the kth element of w.

$$J(k) = w_k^2$$

In all the iterations of the RFE, a linear SVM model is trained and the feature with the smallest weight value has been recognized and eliminated, since it has the least effect on classification. One feature is eliminated in each iteration and remaining features are fed to the SVM model in the next iteration. This feature elimination method has been carried out in a repetitive manner until all the features are eliminated. In a subsequent stage, the feature that has been eliminated last is given more importance. However, this process of feature selection is very time-consuming for a higher dimensional feature vector. To overcome this problem, the elimination of more than one feature in one iteration has been introduced. Considering the highly correlated features, a feature group has been constructed to eliminate more than one feature in a single iteration. In the feature elimination stage, a parameter (elimination threshold value) has been selected in such a way that the features will be eliminated as a correlated feature group when the number of features are more than that predefined parameter value. To achieve more precision, when the number of remaining features becomes same as that elimination threshold value, the RFE method removes single feature in each iteration. Introduction of this correlated feature elimination technique not only reduces the time consumption for feature selection process but also helps to identify the important features by eliminating correlated features present in a large feature set. This approach is generic in nature. It can be applied for other topologically similar problems. Algorithm 4.1 has elaborated the feature selection method.

Algorithm 4.1:
SVM-RFE feature selection [19]

Input: A set of training samples with feature dimension $\boldsymbol{d}$;
A nonlinear SVM training algorithm; T_e.
1: Initialize the list of existing features $F_{in} \leftarrow 1, \ldots, d$;
the list of eliminated features $F_{out} \leftarrow \varnothing$;
2: **while** $F_{in} \neq \varnothing$ **do**
3: Train an SVM model with the features in F_{in}.
4: Calculate the features' ranking criteria
5: Sort F_{in} according to descending order of the ranking criteria.
6: **if** $|F_{in}| > T_e$ **then**
7: $r = \min\left(floor\left(\frac{|F_{in}|}{2}\right), |F_{in}| - T_e\right)$,
8: else
9: $r = 1$.
10: end *if*
11: $F_{removed} \leftarrow the \quad last \quad r \quad elements \quad in \quad F_{in}$;
$F_{in} \leftarrow the \quad first \quad |F_{in}| - r \quad elements \quad in \quad F_{in}$.
12: $F_{out} \leftarrow [F_{removed}, F_{out}]$.
13: **end while**
Output: A ranked list of features $F_{ranked} = F_{out}$, the most important feature in the first place.

4.4.2 SVM-RFE with correlation bias reduction technique

Prior to classification, the feature selection method is a very important step that helps to identify the most appropriate and distinguishable features demarcating the target classes from the large number of features. An efficient feature selection algorithm not only increases the classification accuracy but also reduces the computation burden by reducing the data size. Among the several feature selection algorithms proposed in literature, the RFE method combined with nonlinear SVM has recently been introduced and has been found to be effective in various applications. RFE is a weight-based backward elimination or selection algorithm introduced by Guyon et al. [19]. The essence of the RFE method is to train the SVM classifier model with all the features and subsequently to identify those that cause least decrement of the margin between the target classes. In this reported work, SVM-RFE algorithm combined with a CBR method [23] has been used for feature selection. The CBR is introduced to eliminate the problems associated with SVM-RFE, when a large number of highly correlated data exists.

4.4.3 Nonlinear SVM-RFE technique

In the RFE algorithm, the SVM model has first been trained with the entire candidate feature set and all those features have subsequently been ranked according to the order of their estimated weight value. In each iteration, the feature having smallest ranking criterion has been identified and removed, since it has minimum effect on classification. The remaining features have been subjected to the next iteration. Thus eliminating the features in a recursive manner, they have been sorted in a way such that the feature eliminated later has been assigned higher ranking. For a large feature set, the elimination and selection of features in recursive manner is very time-consuming and also introduces correlation bias problem. To deal with the problem of overfitting, the linear SVM-RFE appears to be more suitable feature selection method when the number of samples is very less in comparison to the several thousands of features. As the nonlinear SVM-RFE can fit the data with less bias, hence, it is expected to outperform the linear one for larger sample size (more than 100). Nonlinear SVM maps the entire features into a higher dimension space. For a given set of training samples $\{x_i, y_i\}$, $x_i \in R^d$, $y_i \in \{-1, +1\}$, $i = 1, 2, \ldots, n$ the nonlinear SVM maps the feature set into a higher dimensional space:

$$x \in R^d \mapsto \phi(x) \in R^h$$

In the new higher dimension feature space the samples are expected to linearly separable. Thus the Lagrangian formulation of nonlinear SVM can be written as:

$$L_D = \sum_{i=1}^{n} \alpha_i - \frac{1}{2} \sum_{i,j=1}^{n} \alpha_i \alpha_j y_i y_j \phi(x_i) \cdot \phi(x_j)$$

where α_i are the Lagrange multipliers. Here we can replace $\phi(x_i)\,.\,\phi(x_j)$ with a kernel function $K(x_i, y_j)$ without the information regarding ϕ as the only form that $\phi(x)$'s have been involved in the training algorithm is their inner product. Among several kernel functions, the radial basis function (RBF kernel) is the common choice, given as

$K(x_i, y_j) = e^{-(||x_i - x_j||^2/2\sigma^2)}$ $||x_i - x_j||^2$ may be considered as squared Euclidean distance between two feature vectors. The above equation can be simplified by replacing $\gamma = (1/2\sigma^2)$ as,

$$K(x_i, y_j) = e^{-\gamma||x_i - y_j||^2}$$

Since the form of ϕ is unknown, the weight vector cannot be obtained. The ranking criterion for a feature k can be written as:

$$J(k) = \frac{1}{2}\sum_{i,j=1}^{n} \alpha_i\alpha_j y_i y_j K(x_i, x_j) - \frac{1}{2}\sum_{i,j=1}^{n} \alpha_i\alpha_j y_i y_j K(x_i^{(-k)}, y_j^{(-k)})$$

The notation $(-k)$ indicates the removed kth feature. The feature with smallest J's will be recursively removed from the feature set in iteration of RFE.

4.4.4 Correlation bias reduction

In the classification of high-dimensional data containing groups of correlated features, the assessment of the appropriacy of the feature is of utmost importance. To achieve this, the model that gives priority to the retrieval of all predictive features, should be preferred. Toloşi et al. [24], appraised that the features in the correlated feature groups received smaller weights due to the partaken responsibility in the classification models. These smaller weights will act against the undermining of the importance of those features that are highly relevant. The elimination of features because of the incorrect underestimation of their ranking criterion is known as "correlation bias." To reduce the correlation bias, the most acceptable strategy is to group the correlated features prior to the model fitting and identify the feature representative of each group which defines the importance of the original features. SVM-RFE feature selection method eliminates one feature in each iteration according to their ranking criterion. Thus this method is not efficient for higher dimension features. To overcome this issue, a group of features are eliminated in one iteration. This may lead to the removal of a group of correlated as well as relevant features entirely. Introduction of the CBR technique move back the representative feature from the eliminated features set to the surviving features set. In each iteration, the group representative feature has been selected as the feature with the highest ranking criterion (weight value). This technique monitors and corrects the wrongly estimated weight and rank of features due to correlation bias, without changing ranking criterion of the original feature set. In different feature selection algorithms including SVM-RFE, highly correlated features bring wrong estimations of the weight and rank of the features [14].

In the preliminary stage of the SVM-RFE with CBR feature selection algorithm, features have been eliminated recursively according to their ranking criterion. A predefined threshold value has been selected as the elimination threshold. When the number of features in the existing feature set is more than the predefined elimination threshold value, the RFE method will eliminate a batch of features in each iteration. After reaching at the predefined threshold value, a single feature with lowest ranking criterion will be eliminated in the remaining iterations. The purpose of the inclusion of CBR algorithm with SVM-RFE method is to move the potentially useful features from eliminated features set to surviving features set. In order to identify highly correlated feature groups, two threshold values, correlation threshold and group threshold, have been selected. In the eliminated feature set, the bunch of features (more than group threshold value) having an absolute correlation coefficient higher than the predefined correlation threshold value are considered as a group. If none of the group members are present in the existing feature set, then the feature with highest ranking criterion has been moved to the existing feature set as the group representative. This operation has been continued for all features in the eliminated features set.

4.5 Application of skin disease classification techniques

4.5.1 Classification of malignant melanoma and benign nevi using automatic correlation bias reduction technique

The noninvasive computerized image analysis techniques have a great impact on accurate and uniform evaluation of skin abnormalities. In this section, the work reports a method for the differentiation of common melanoma from benign nevi based on texture and morphological features extracted from dermoscopic images. For the identification of two disease classes, a 2D wavelet packet decomposition based fractal texture analysis has been used to extract the irregular texture pattern of the skin lesion area. On the whole 6214 features have been extracted from each of the 4094 skin lesion images, by analyzing the textural pattern and morphological structure of the lesion area. For the identification of the most efficient feature set, an improved CBR method has been introduced in combination with SVM-RFE. An automatic selection of correlation threshold value has been introduced in this work to eliminate the correlation bias problem associated with SVM-RFE algorithm. With these selected features, the SVM classifier with RBF is employed.

4.5.2 Materials

This study has been done on 4094 skin lesion images of two different classes, namely common melanoma and benign nevi diseases. The skin lesion images have been collected from some widely acceptable image databases [25–28].

4.5.3 Methodology

In this study, the dermoscopic images of melanoma and nevus disease classes have been considered. Skin lesion area have been segmented from the dermoscopic images employing morphological segmentation technique as discussed in Chapter 2, Preprocessing and segmentation of skin lesion images. Morphological features and lesion border irregularity have been estimated from the segmented images using statistical methods and wavelet fractal based border irregularity measurement technique (discussed in Chapter 3: Extraction of effective hand crafted features from dermoscopic images). Using these feature extraction techniques, overall 6214 number of features have been extracted from each of the dermoscopic images. For the differentiation of the two disease classes, a binary SVM classifier has been chosen. To obtain an improved classification performance, selection of most differentiating and demarcating features is important. Prior to the classification model, a RFE based feature selection technique is introduced. However, in this wide range of feature vector, a much higher probability exists for the presence of correlated features. Considering this condition, CBR technique is incorporated with the RFE methodology. Therefore to improve the classification performance using large feature set SVM-RFE with CBR technique is implemented. As discussed in previous section, the selection of correlation threshold value is a very critical parameter. Yan and Zhang [23] has elaborated how different classification accuracies have been obtained for various settings of correlation threshold value. The correlation threshold value helps to identify the correlated features group. For different correlation threshold values, the representative feature of a group may different correlation threshold value is usually reflected in the correct classification accuracy. In this section, to eliminate the effect of the correlation threshold value on the classification accuracy, the said threshold values has been varied to yield more and more improved classification accuracy. A wide range of correlation value has been selected from 0.5 to1. For each of the selected threshold value, the features have been ranked according to their weight values using SVM-RFE with CBR technique. After the feature selection step, the images have been classified using SVM classifier with RBF. The performance of classification has been evaluated by calculating some standard parameters like the sensitivity (SN), the specificity (SP) and the accuracy (ACC). The correlation threshold value which leads to the solution with the best classification accuracy, has been considered as final. For the selection of important and distinguishable features between the two classes of skin lesion images using SVM-RFE with CBR technique, three guiding parameters have been chosen as the group threshold value, the correlation threshold value and the elimination threshold value. Among these three parameters, the correlation threshold value has been selected automatically according to the best classification accuracy. A larger group threshold values identifies fewer group of features and hence eliminates some groups of correlated features too early [23]. As a result, the classification accuracy is also degraded. Taking into consideration the effect of the group threshold value on classification performance, the entire feature selection has been done with a predefined group threshold value as 1. The elimination threshold value has been selected by rigorous exercise, as

60 features. When the number of existing features is more than 60, then half of the existing features will be moved to the eliminated feature set in each iteration. When the number of features becomes 60, a single feature will be eliminated in each iteration.

4.5.4 Results and discussions

Using the SVM-RFE with CBR feature selection algorithm, the features that are most conspicuous in revealing the demarcation between melanoma and dysplastic nevi, have been selected according to their ranking criterion. Selecting the first 20, 30, 40, and 50 features at each stage, the images have been classified using SVM classifier with RBF kernel using tenfold cross-validation method. The classification performance has been evaluated by calculating the SN, the SP and the ACC attained during the classification stage. In this presented work, the performance of the classification of common melanoma and benign nevi has been assessed for each of the selected mother wavelet. The classification performance has been evaluated by calculating sensitivity, specificity and accuracy for Daubechies, Symlet and Coiflet wavelet families, for different number of selected features. Prior to the classification, SVM-RFE with CBR feature selection algorithm has been used to rank the features according to their weight values and the images have been classified for first 20, 30, 40, and 50 selected feature set. The entire exercise has been carried out using different feature sets, selected using varying correlation threshold values from 0.5 to1. At maximum correlation, that is, 1, the feature selection has been done using SVM-RFE, without introducing the CBR technique. In this study, very good and consistent correct identification accuracy has been achieved for most of the classifiers with a minimum accuracy of 85.05% for Daubechies-6 wavelet function with first 50 features. From the computations carried out, it has been found that better classification performance has been obtained for the lower order wavelet functions. With an efficient feature selection method, it is expected to classify the images with acceptable performance metrics using lesser number of features. In this work, the classification performance with first 30 features have been considered as the performance baseline. A minimum correct classification accuracy as 91.18% for first-order and third-order Coiflet mother wavelet and maximum accuracy as 98.28% for third-order Daubechies wavelet have been achieved by considering first 30 selected features. For Daubechies wavelet family, the maximum sensitivity, specificity and accuracy values have been achieved as 97.63%, 100%, and 98.28%, respectively, with 3rd order mother wavelet. The highest sensitivity, specificity and correct classification accuracy have been achieved as 97.62%, 100%, and 98.25%, respectively, for Symlet wavelet family. Similarly, 95.99% sensitivity, 99.08% specificity and 96.81% correct classification accuracy have been obtained for second-order Coiflet wavelet using first 30 selected features.

The first five important features for third-order Daubechies, seventh-order Symlet, and second-order Coiflet wavelet functions have been given in Table 4.1. From

Table 4.1 First five selected features using SVM-RFE with CBR technique.

Mother wavelet	First feature	Second feature	Third feature	Fourth feature	Fifth feature
Daubechies-3	Tex: FD_H1.2_B	BI: HFD_D1.2	Tex: Ent_0_B	Tex: FD_V1.2.2_B	Morph: Minor_L
Symlet-7	Tex: FD_D1.1_B	Tex: FD_V1.2.1_B	Tex: FD_V1.2.2_B	Tex: Mean_H1.2_G	Morph: Minor_L
Coiflet-2	Tex: Var_H1.2.4_B	Morph: Minor_L	Tex: FD_V1.2.2_B	Tex: Mean_H1.2_G	Tex: FD_D1.2.3_B

Tex: texture features; *Morph*: morphological features; *FD*: fractal descriptor; *Ent*: entropy; *Minor_L*: minor axis length; *HFD*: Higuchi fractal dimension; *Mean*: mean intensity; *Var*: variance; *B*: blue channel of RGB color image; *G*: green channel of RGB color image; *SVM-RFE*, support vector machine-based recursive feature elimination; *CBR*, correlation bias reduction.

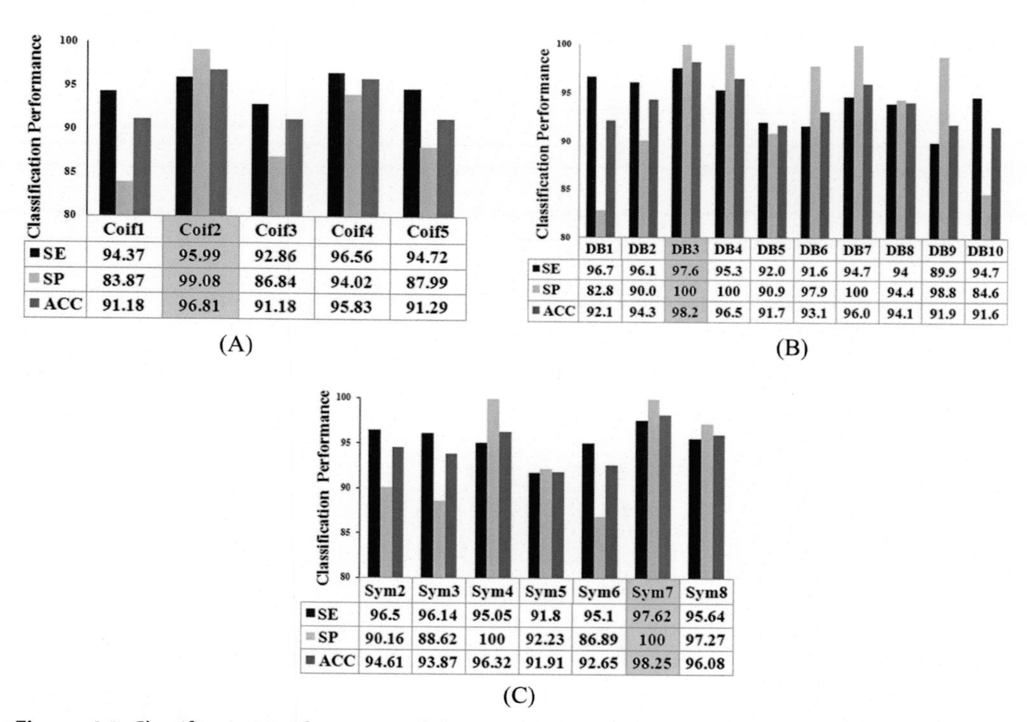

Figure 4.1 Classification performance of the proposed methodology using different wavelet families (A) Daubechies, (B) Symlets, and (C) Coiflets for top 30 selected features using support vector machine-based recursive feature elimination with correlation bias reduction technique. *ACC*, Accuracy; *SE*, sensitivity; *SP*, specificity.

Table 4.1 it can be seen that the fractal descriptor and statistical features related to the texture, are the most dominating features for all three types of mother wavelets considered. The classification performance for Daubechies, Symlet, and Coiflet wavelets for the first 30 selected features using SVM-RFE with CBR technique, have been shown in Fig. 4.1. In Fig. 4.2, the effect of the number of selected features on the classification performance have been shown by representing the variation of classification performance parameters (sensitivity, specificity and accuracy) with the number of features from 20 to 50. It can also be clearly seen that, as expected, increase in the number of features improves the classification performance. However, for the sake of achieving highly acceptable result with lesser number of features, first 30 features have been considered. In the feature selection stage using SVM-RFE with CBR, the selection ofcorrelation threshold value is of prime importance. In Fig. 4.3, the variation of the classification accuracy with the correlation threshold value, has been shown for three selected mother wavelets. From Fig. 4.3, it has been observed that for the correlation threshold value ranging from 0.75 to 0.9 the highly acceptable classification

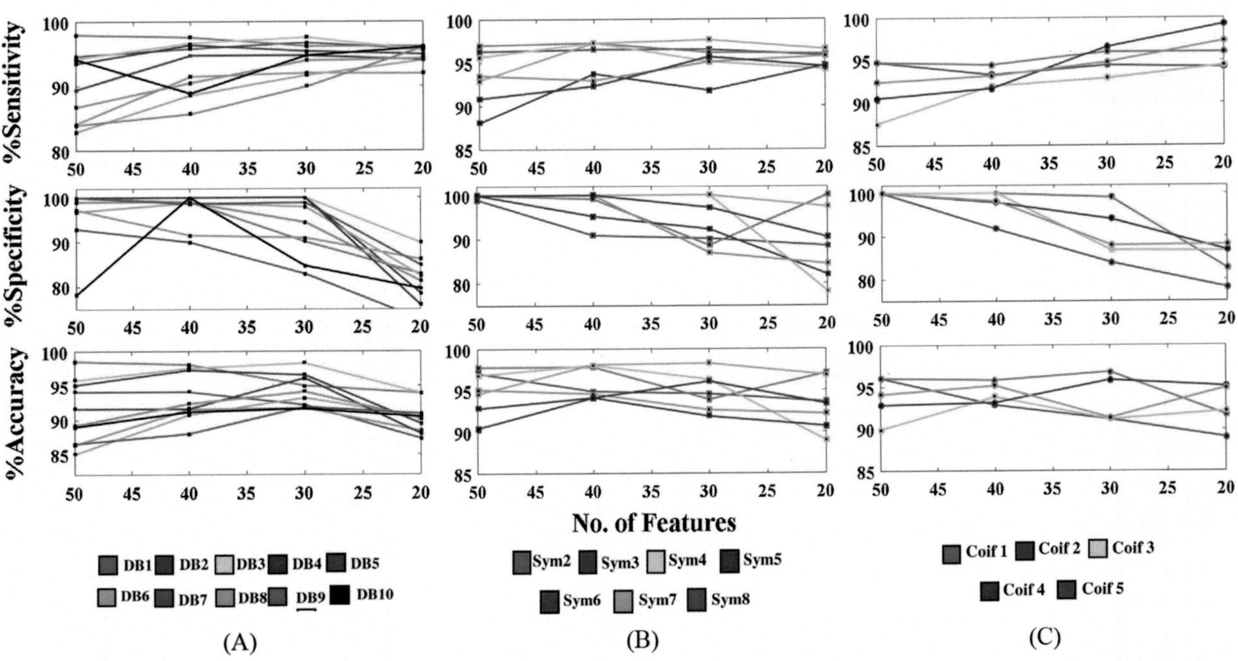

Figure 4.2 Variations of the classification performance indices according to the number of selected features for (A) Daubechies, (B) Symlet, and (C) Coiflet wavelet families.

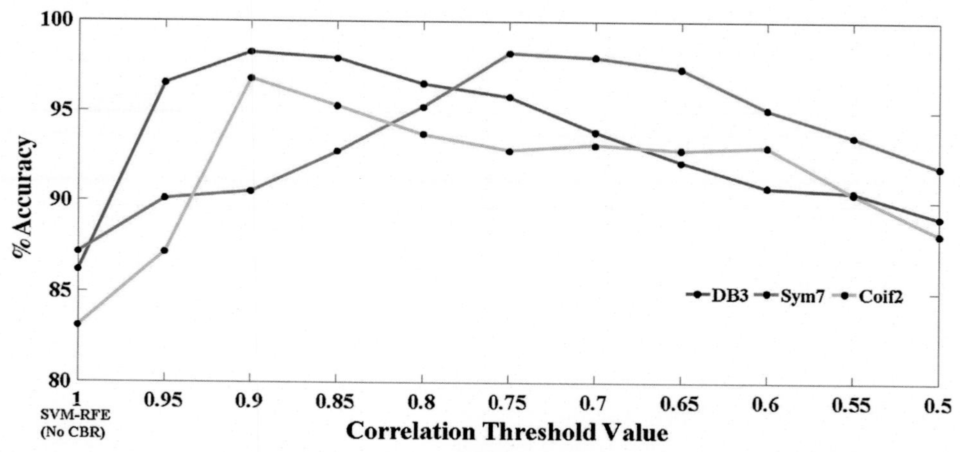

Figure 4.3 Change of correct classification accuracy according to the correlation threshold value for third-order Daubechis, seventh-order Symlet, and second-order Coiflet mother wavelet. *SVM-RFE*, Support vector machine-based recursive feature elimination; *CBR*, correlation bias reduction.

Table 4.2 Comparison on classification performance for different wavelet functions using SVM-RFE and SVM-RFE with CBR technique.

Mother wavelet	SVM-RFE			SVM-RFE with CBR		
	SE	SP	ACC	SE	SP	ACC
Daubechies-3	91.25	80.31	86.16	97.63	100	98.28
Symlet-7	92.11	84.23	87.19	97.62	100	98.25
Coiflet-2	89.63	78.67	83.12	95.99	99.08	96.81

ACC, Accuracy; *CBR*, correlation bias reduction; *SE*, sensitivity; *SP*, specificity; *SVM-RFE*, support vector machine-based recursive feature elimination.

accuracy has been obtained. For Daubechies wavelet family, the highest accuracy has been obtained as 98.28% for 0.9 correlation threshold value. Similarly the correlation threshold values corresponding to the highest accuracy for Symlet and Coiflet wavelet family (98.25% and 96.81%, respectively) have been obtained as 0.75 and 0.9, respectively. The results of a comparative study between SVM-RFE and SVM-RFE with CBR feature selection technique have been tabulated in Table 4.2. From this table, it can be inferred that with regard to the classification performance, the introduction of the modified CBR technique with SVM-RFE feature selection method has outperformed the SVM-RFE method remarkably. A comparative study of the proposed methodology with other state-of-the-art techniques has been presented in Table 4.3. Oliveira et al. [7] have achieved 74.36% classification accuracy for the identification of nevus and melanoma images by introducing fractaldimension measurement technique of the lesion area, statistical color features extraction and border irregularity

Table 4.3 Performance comparison of the proposed method with state-of-the-art techniques.

Work	Database	No. of images	Classification performance (% Accuracy)
Oliveira et al. [7]	Loyola University Chicago, YSP Dermatology Image Database, DermAtlas, DermIS, Saúde Total, Skin Cancer Guide, Dermnet—Skin Disease Atlas	408	74.36
Kasmi and Mokrani [29]	EDRA, Interactive Atlas of Dermoscopy	200	94
Shimizu et al. [3]	Keio University Hospital, University of Naples and Graz, Tokyo Women's Medical University	964	90.48
Abuzaghleh et al. [30]	PH2 database	200	97.50
Rastgoo et al. [4]	Vienna General Hospital	5130	Sensitivity—98% Specificity—70%
Barata et al. [5]	Hospital Pedro Hispano, Matosinhos	176	Sensitivity—96% Specificity—80%
Garnavi et al. [31]	–	289	91.26
Proposed method	International Dermoscopic Society, Dermoscopic Atlas, ISIC: challenge 2017, and PH2 database	4094	Sensitivity—97.63% Specificity—100% Accuracy—98.28%

measurement using inflexion point and vector product descriptor method. In [29] Kasmi et al. have implemented the automatic ABCD rule to differentiate malignant melanoma from benign skin lesions and have reported a classification accuracy of 94%. The shape, color and asymmetry of brightness have been measured using different mathematical operations along the smaller subregions of the original image. The color features have been extracted by measuring the normalized Euclidean distances between each pixel of the lesion and six suspicious colors (white, black, red, light-brown, dark-brown, and blue–gray) along with the presence of different structures in the lesion area. Shimizu et al. [3] have reported 90.48% accuracy for melanoma identification using layered model. In [30], Abuzaghleh et al. proposed a noninvasive real-time skin lesion identification technique by extracting 2D Fast Fourier Transform, Discrete Cosine Transform and color histogram features with an accuracy of 97.50% for melanoma identification. The sensitivity of 98% has been obtained by Rastgoo et al. [4] for the differentiation of melanoma from dysplastic nevi by calculating the distance and color variance with complete local binary pattern, gray-level co-occurrence matrix (GLCM), histogram of oriented gradient features. The local gradient and color histogram features have been extracted for the detection of melanoma images with the 96% sensitivity and 80% specificity by Barata et al. [5]. Garnavi et al. [31] have achieved 91.26% classification accuracy for the diagnosis of melanoma, measuring different irregularity indices and statistical features at each level of the four level wavelet tree. The tabulations reveal that this study achieves decent classification performance compared to the recently published methods. From Table 4.3, a conclusion can be drawn that this methodology extracts the representative features and classify the melanoma using SVM classifier incorporating modified SVM-RFE with CBR method, with a highest accuracy level.

4.6 Three-class classification of skin lesions using recursive feature elimination based feature selection technique

In this study, the focus has been on the identification of each of the three critical and characteristically close skin diseases, namely melanoma, dysplastic nevi, and BCC, accurately. Close resemblance of the visual characteristics between melanoma and nevus throws a challenge to the dermatologist for accurate identification. BCC is also regarded as one type of malignant epidermal lesion thereby can be erroneously identified as malignant melanoma. Some of those closely similar lesion images that are difficult to identify by visual inspection, have been shown in Fig. 4.4. This identification has been attempted by extracting the shape, texture, and color-related features from the dermoscopic images of various skin lesions. Border asymmetry is one of the key features of melanocytic skin lesion. Here, the degree of irregularity of lesion border has been quantified by estimating the fractal dimension. For the texture feature extraction along with popular GLCM method, a fractal based regional texture analysis (FRTA) algorithm has been introduced to extract the textural irregularity of the lesion area. Along with the shape and the texture features, different statistical measures have been extracted from each of the color channels of RGB, *CIEl_a_b* and HSV color dermoscopic images. To select the effective features from this large feature set, SVM-RFE algorithm has been used. Here, the primary concentration has been given to the development of an integrated system comprised of image preprocessing, segmentation followed by feature extraction and efficient multiclass classification of skin diseases. For the three-class classification purpose, a layered structure has been introduced for the differentiation of melanoma, dysplastic nevi and BCC. Apart from the disease identification, this layered structure of classifiers will separate out those images, which have been misclassified in any of the classification layer. The schematic representation of the entire methodology is shown in Fig. 4.5.

4.6.1 Materials

In this study, primarily 6579 number of digital dermoscopic images, consisting of 2419 melanomas, 3261 dysplastic nevi and remaining 899 BCC have been composed from the available datasets in the Internet, [25–28]. The expert dermatologists have histopathologically confirmed the images considered for analysis. The dataset used in this

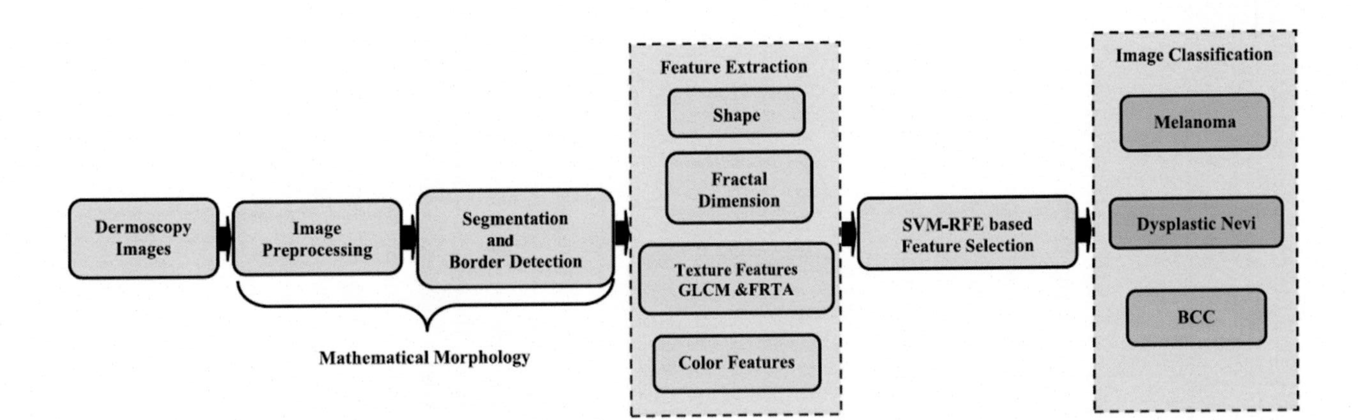

Figure 4.4 Block diagram of the present methodology. *BCC*, Basal cell carcinoma; *FRTA*, fractal based regional texture analysis; *GLCM*, gray-level co-occurrence matrix; *SVM-RFE*, support vector machine-based recursive feature elimination.

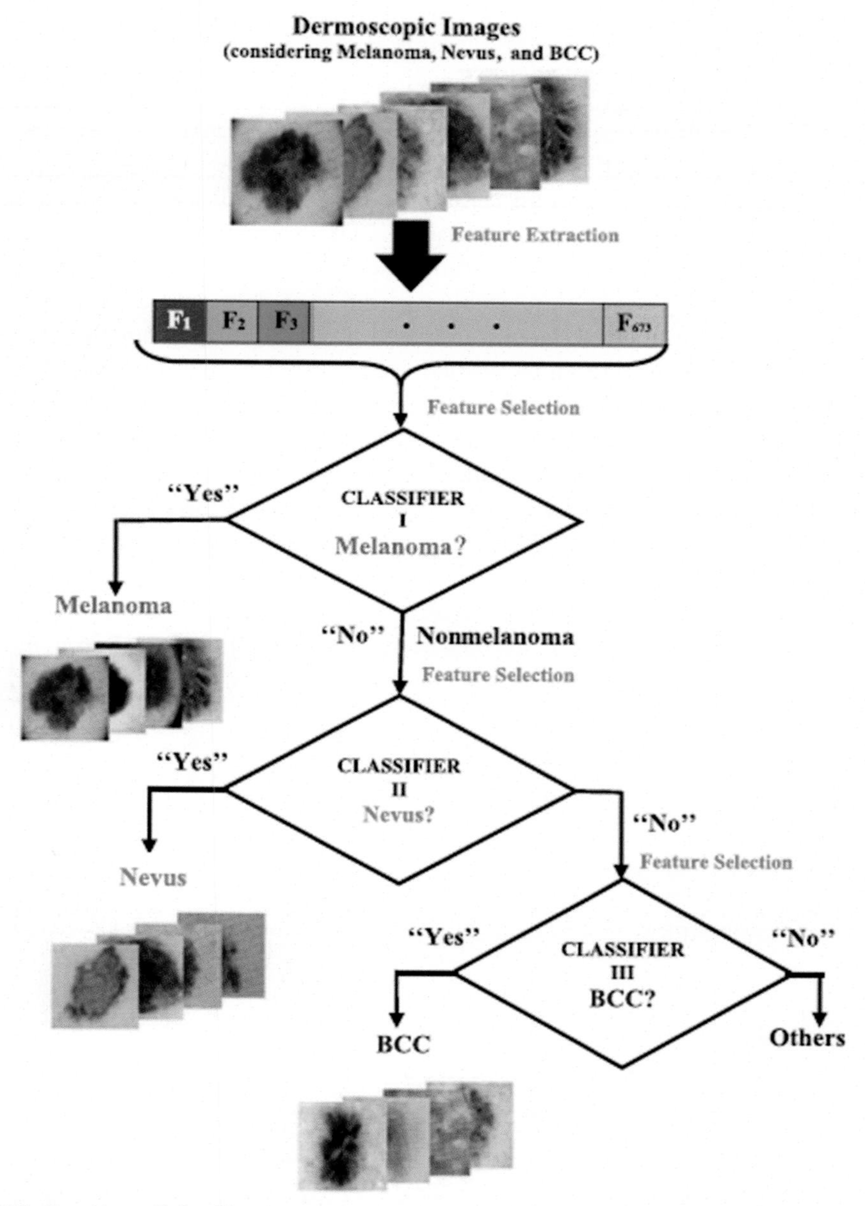

Figure 4.5 Overview of classification strategy using layered structure. *BCC*, Basal cell carcinoma.

study has been elaborated in Chapter 3, Extraction of effective hand crafted features from dermoscopic images.

4.6.2 Methodology

Literature suggests different classification strategies for various biomedical applications [32–34]. In this study, a multiclass classification technique is used for the segregation of three different classes of skin abnormalities. To ensure the correct classification in each stage of the classifier, the classification accuracy must be very high at the first stage. In one-vs-all classification technique different hyperplanes are constructed, depending on the number of classes considered for classification. But this classification model is unable to segregate miss-classified images of each class. In biomedical condition monitoring identification of miss-classified images is very much important for further analysis. Considering this important aspect of biomedical applications, a layered model is introduced here. In this work, a layered structure of classifier that has been shown in Fig. 4.6, is introduced for the classification of the three skin diseases. In this classification strategy, the "CLASSIFIER I" has been dedicated to segregate melanoma lesions from all other skin diseases by considering melanoma as positive class while other diseases have been grouped as negative class of images. Before the classification, the features that highly differentiate melanoma from all other skin abnormalities have

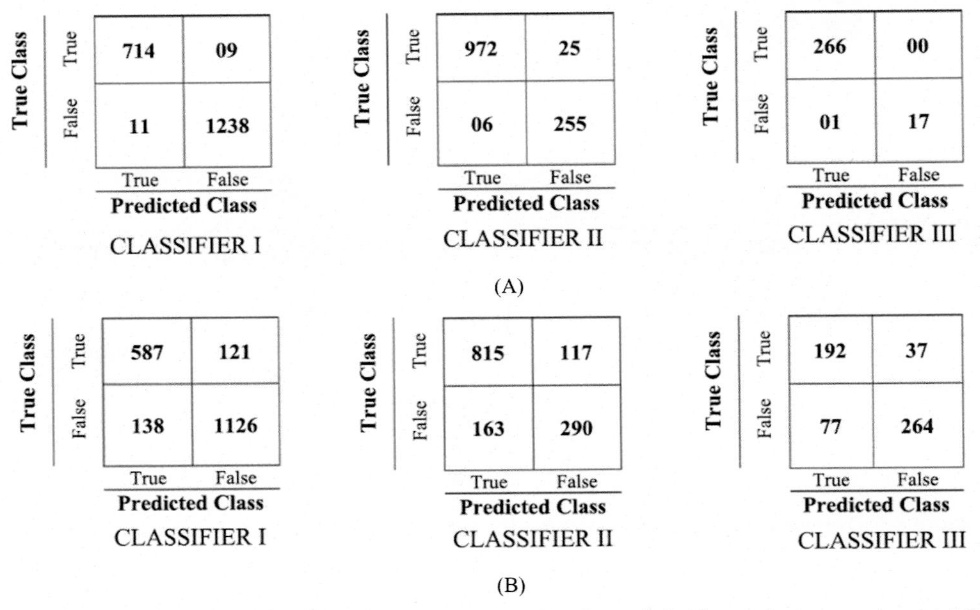

Figure 4.6 Confusion matrices of each of the three classifiers of the layered structured model for first 30 selected features (A) with support vector machine-based recursive feature elimination (SVM-RFE) and (B) without SVM-RFE feature selection method.

been selected using SVM-RFE feature selection method. In the second stage of the layered structure, another classifier as the "CLASSIFIER II" has been used to separate dysplastic nevi from BCC. However, in this stage of binary classification, the images have been grouped as dysplastic nevi and "others," where "others" have the probability of inclusion of the miss-classified melanoma, the misclassified dysplastic nevi and the BCC images. Prior to the classification, another SVM-RFE stage has been introduced to increase the classification accuracy of dysplastic nevi from BCC and other misclassified images (grouped as others). Finally, another binary classifier, the "CLASSIFIER III" has been included in the layered structure for the segregation of BCC from the other misclassified images from each stage of the classification, grouped as "unidentified" or "others."

4.6.3 Results and discussions

The SVM-RFE feature selection algorithm has selected the important and demarcating features according to their ranking criterion for the improvement of the classification performance. Table 4.4 shows the top three selected features for each of the classification stage. The kernel function and associated kernel hyperparameters of SVM classifier have been determined by evaluating the classification performance indices. The kernel hyperparameters, penalty parameter of soft margin cost function and the scale parameter of the kernel have been estimated using grid-search-based algorithm. For the selection of kernel function, three types of kernel functions namely, RBF, polynomial kernel (second and third order) have been chosen. Considering each of the kernel functions, the classification performance indices (SN, SP, ACC) have been determined for each of the classifiers. It is clear from the entries in Table 4.5 that highly acceptable results have been obtained using SVM classifier with RBF kernel, compared to the second- and third-order RBF kernels. The classification performance

Table 4.4 First three selected features using support vector machine-based recursive feature elimination feature selection method for each classifier.

	Classifier	First feature	Second feature	Third feature
Layer structure	"CLASSIFIER I" (melanoma vs nonmelanoma)	Col: Ent_G[#] (Entropy)	GLCM: Eng (Energy)	Morph: FD[$] (Fractal Dim)
	"CLASSIFIER II" (dysplastic nevi vs nonnevus)	Col: Ent_H* (Entropy Hue)	Col: Var_L[†] (Variance Lightness)	GLCM: Max (Maximum)
	"CLASSIFIER III" (BCC vs others)	MSFTA: FD_3 (HFD)	Col: Mean_H (Mean Hue)	GLCM: Autoc (Autocorrelation)

Feature categories: Col: Color features; Morph: morphological features; FD_3: Haussdorf fractal dimension (HFD) of third binary image region.
[#]G channel of RGB color image.
[$]Fractal dimension.
*H channel of HSV color image.
[†]L channel of CIEl*a*b color image.

Table 4.5 Comparative study on classification performance for the support vector machine classifiers with different kernel functions.

Kernel	CLASSIFIER I			CLASSIFIER II			CLASSIFIER III		
	% SE	% SP	% ACC	% SE	% SP	% ACC	% SE	% SP	% ACC
RBF	98.48	98.28	98.99	99.39	91.07	97.54	99.63	100	99.65
Polynomial (2nd order)	96.13	97.01	96.52	97.26	97.00	97.58	97.86	97.67	97.81
Polynomial (3rd order)	97.31	98.06	97.92	98.11	98.37	98.41	98.07	98.69	98.42

ACC, Accuracy; *SE*, sensitivity; *SP*, specificity.

of each stage of the classifier has been summarized based on three classification baseline parameters, namely SN, SP, and correct classification ACC as the following:

SE = No. of true positive (TP) skin disease samples
/No. of all skin disease samples classified as positive (TP + FN)

SP = No. of true negative (TN) skin disease samples
/No. of all skin disease samples classified as negative (TN + FP)

ACC = No. of correctly classified skin disease samples (TP + TN)
/Total no. of skin disease samples (TP + TN + FP + FN)

where TP, positive samples classified as positive; TN, negative samples classified as negative; FP, negative samples misclassified as positive; FN, positive samples misclassified as negative.

To classify the three disease classes using this layered structure model, the number of training samples have been gradually decreased in each layer. Classification of the melanoma images in the first layer classifier "CLASSIFIER I" with a higher sensitivity ensures lesser number of misclassified images in the next classification stage. So, the better classification performance at the higher stage guarantees reduced variability in the feature set for the subsequent stages. The impact of classification performance of the first stage on the performances of the "CLASSIFIER II" and "CLASSIFIER III" can be well understood from Table 4.5.

The classification of the skin diseases from the dermoscopic images has been done using SVM classifier with RBF kernel and the classification performance has been measured by dividing the entire dataset into training and testing samples. The classification accuracy has been evaluated by gradually decreasing the training samples from 80% to 50% of the entire dataset. The classification performance for each of the training and testing sets for all the three classifiers have been given in Table 4.6. The first column of Table 4.6 denotes the classification stage of the layered structured classification strategy. From Table 4.6, it has been observed that with the decrement in the number of training samples, the classification performance indices have decreased.

Table 4.6 Classification performance (in %) of layer structure using support vector machine classifier (RBF kernel) with SVM-RFE of varying size of training and testing samples.

Classifier	Training/testing samples (%)	SE	SP	ACC
CLASSIFIER I	80/20	99.44	99.43	99.14
	70/30	98.48	98.28	98.99
	60/40	89.65	94.26	92.13
	50/50	81.47	88.61	87.27
CLASSIFIER II	80/20	99.49	92.14	97.93
	70/30	99.39	91.07	97.54
	60/40	90.03	94.22	93.32
	50/50	78.69	83.04	81.16
CLASSIFIER III	80/20	99.63	88.89	99.30
	70/30	99.63	100	99.65
	60/40	87.17	92.91	91.26
	50/50	78.57	71.37	74.83

ACC, Accuracy; *SE*, sensitivity; *SP*, specificity; *SVM-RFE*, support vector machine-based recursive feature elimination.

Among 80% and 20% training and testing samples of the entire dataset, the layered structure has achieved higher classification performance. Considering a moderate number of training and testing samples with proper balance and acceptable results in each classification stages for the classification of the melanoma, nevus, and BCC images, the training dataset has been constructed by considering 70% of images from each class (considering all possible varieties depending on the clinical features). Among 6579 number of digital dermoscopic images, 4607 number of images (1694 melanoma, 2283 nevus and 630 BCC) have been used in training phase and remaining 1972 number of images in testing phase. To reduce the data imbalance and overfitting problem of the classifier, the data augmentation techniques have been employed on the 70% training sample images. The application of data augmentation techniques in the training samples ensures that the same sample has not been used in both the training and testing phase.

In Table 4.7, the number of features have been determined using SVM-RFE feature selection technique [19], according to their ranking for each of the classifier. According to their ranking criterion, the first 20, 30, 40 and 50 number of features have been selected and fed to the classifier for the identification of diseases. For the classifier without prior SVM-RFE feature selection method, the number of features can be considered to be randomly selected from the entire feature set. The classification performance has been measured at each stage and depending on the classification accuracy, the number of features has been varied with an increment of 10 features according to their ranking criteria in SVM-RFE as shown in Table 4.7. The corresponding confusion matrices have been shown in Fig. 4.7. The performance of the

Table 4.7 Classification performance (in %) of layer structure using SVM classifier (RBF kernel) with and without RFE of varying feature size.

	SVM (RBF kernel) with RFE				SVM (RBF kernel) without RFE		
Classifier	**#No. of features**	**SE**	**SP**	**ACC**	**SE**	**SP**	**ACC**
CLASSIFIER I	20	96.24	97.39	96.99	76.45	89.83	80.35
	30	98.48	98.28	98.99	80.97	90.30	86.17
	40	99.07	99.39	99.28	79.52	89.65	85.89
	50	99.44	99.79	99.67	82.81	91.03	89.69
CLASSIFIER II	20	99.22	74.73	96.99	79.95	73.36	78.07
	30	99.39	91.07	97.54	83.33	71.25	79.78
	40	99.57	91.07	99.09	79.97	75.34	78.60
	50	100	96.23	99.79	85.73	81.66	82.36
CLASSIFIER III	20	95.92	95.24	95.60	52.63	85.16	75.14
	30	99.63	100	99.65	71.38	87.70	80.00
	40	100	100	100	70.18	92.24	84.97
	50	100	100	100	81.13	92.91	89.64

ACC, Accuracy; *RBF*, radial basis function; *RFE*, recursive feature elimination; *SE*, sensitivity; *SP*, specificity; *SVM*, support vector machine.

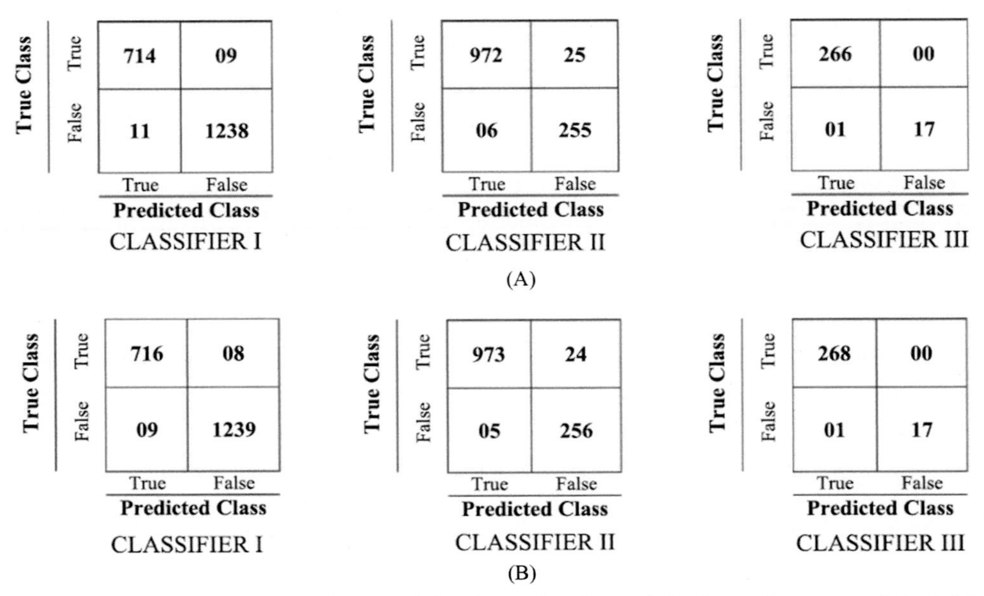

Figure 4.7 Confusion matrices of each of the three classifiers of the layered structured model for first 30 selected features (A) with linear support vector machine-based recursive feature elimination (SVM-RFE) and (B) nonlinear SVM-RFE feature selection method.

proposed multiclass classification strategy has been evaluated with linear as well as nonlinear recursive feature elimination (RFE) feature selection method prior to the classification. The feature selection algorithm has been introduced to improve the classification performance by considering the important and demarcating features from the entire feature set. Here, RFE based feature selection technique has been used. Guyon et al. [19] has introduced the RFE technique implementing weight-based backward feature elimination or selection method. In the RFE algorithm, the SVM classifier has been trained with all the features and subsequently the features that have caused the least decrement in the interclass margin, have been selected. In the SVM-RFE feature selection algorithm, all the features have been ranked according to their weight values, determining their effect in the classification. In each iteration, a bank of features has been eliminated depending on their ranking criterion, until the available number of features is same as a preselected input parameter, namely feature elimination threshold value. As the number of features reaches the threshold parameter value, single feature has been eliminated in each iteration, until all the features have been ranked sequentially. Here, for selecting the appropriate features, the feature elimination threshold value has been determined experimentally as 60, depending on the classification performance.

For each of the classifier with RBF kernel, the SE, the SP and the ACC have achieved highly acceptable values for the first 50 selected features. Table 4.7 also suggests that as the number of selected features has been decreased gradually, the classification performance does not deviate remarkably for the SVM classifiers with RBF kernel. In layered structure with RBF kernel, "CLASSIFIER I" has achieved the sensitivity as 99.44% and 98.48% with a correct classification accuracy as 99.67% and 98.99% for first 50 and 30 features, respectively. To deal with the overfitting problem of the classifier, it is preferable to obtain satisfactory classification performance with a smaller set of features. According to Table 4.5, a highly acceptable performance with an accuracy of 98.99%, 97.54% and 99.65% for "CLASSIFIERI," "CLASSIFIER II" and "CLASSIFIER III," respectively, have been achieved with first 30 selected features. So the classifiers with the first 30 features provide highly acceptable results with good balance at each of the classification stage of the layered structure with more than 90% sensitivity and specificity. For this proposed three-class classification technique using layered structure, SVM classifier without prior feature selection algorithm has also been used for the classification performance evaluation as shown in Table 4.7. In the first stage of the classification, the classification accuracy has been achieved as 89.69% and 86.17% for the randomly selected first 50 and 30 features from the entire extracted feature set. As in this first stage, high identification accuracy has not been achieved, the variability in the feature set has increased in the second stage of the classification due to the inclusion of a large number of misclassified images from the previous stage. As a result, an acceptable classification performance has not been achieved

Table 4.8 Classification performance (in %) of layer structure using SVM classifier (RBF kernel) with Linear and nonlinear RFE method.

		SVM with linear SVM-RFE			SVM with nonlinear SVM-RFE		
Classifier	**#No. of features**	**SE**	**SP**	**ACC**	**SE**	**SP**	**ACC**
CLASSIFIER I	30	98.48	98.28	98.99	98.75	99.36	99.14
CLASSIFIER II	30	99.39	91.07	97.54	99.49	91.43	97.69
CLASSIFIER III	30	99.63	100	99.65	99.63	100	99.65

ACC, Accuracy; *RBF*, radial basis function; *RFE*, recursive feature elimination; *SE*, sensitivity; *SP*, specificity; *SVM*, support vector machine.

by the "CLASSIFIER II." In Table 4.7, the classification performance baseline parameters using SVM classifier with RBF kernel indicate that the considered feature set for the classification of the three classes of skin lesions does not contain significant number of important and demarcating features. The performance comparison of the proposed methodology has been validated from the corresponding confusion matrices for each of the classifier stages, as shown in Fig. 4.6. From the confusion matrices, the number of test images and their identification rate has been observed for with and without SVM-RFE feature selection strategy.

Table 4.8 depicts the comparative performance analysis considering the first 30 features according to their ranking using linear and nonlinear SVM-RFE feature selection method. Prior to the classification, the first 30 ranked features have been selected using linear and nonlinear feature selection methods. From Table 4.8, it is clear that the CLASSIFIER I has identified melanoma images with higher sensitivity for nonlinear SVM-RFE than with the linear SVM-RFE method. As in the first stage of the classification, a higher sensitivity has been achieved, the classification performances have been improved in the subsequent stages. The CLASSIFIER II and CLASSIFIER III have identified the nevus and BCC images with an enhanced sensitivity of 99.49% and 99.63%, respectively. Although, higher classification performance indices have been reported in Table 4.8, the nonlinear SVM-RFE method requires more time than linear SVM-RFE feature selection method. The corresponding confusion matrices have been shown in Fig. 4.7.

The reported performance of the proposed classification scheme can be validated from the confusion matrix as shown in Fig. 4.8. Each element of the confusion matrix represents the probability of the predicted class with respect to the ground truth or true class. From the confusion matrix, it can be observed that the "CLASSIFIER I" and the "CLASSIFIERII" in the layered model have identified melanoma (Class I in Fig. 4.8) and dysplastic nevi (Class II in Fig. 4.8) with a high probability rate. At the final stage of the classification, the "CLASSIFIER III" has been able to identify all the BCC images efficiently.

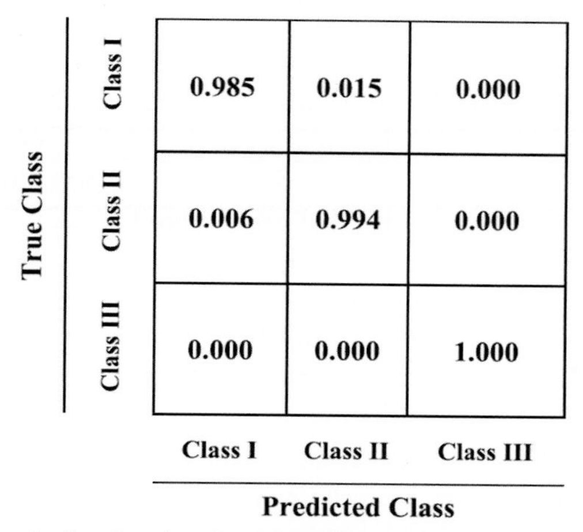

Figure 4.8 Confusion matrix for the classification performance of melanoma, dysplastic nevi, and basal cell carcinoma as classes I, II, and III, respectively.

Table 4.9 Classification performance (in %) comparison of proposed method with other classifiers.

Classifier	Classification performance indices (in %)		
	SE (%M, %N, %BCC)	SP (%M, %N, %BCC)	ACC (%M, %N, %BCC)
KNN ($K=3$) [33]	80.86, 78.89, 78.63	86.23, 81.14, 82.21	85.79, 80.21, 80.88
NB [37]	91.15, 87.65, 89.21	90.08, 88.08, 90.16	90.81, 88.10, 89.97
ANN [36]	94.95, 94.26, 95.17	96.32, 93.13, 97.77	96.06, 94.11, 97.50
Decision tree [36]	88.73, 87.15, 87.69	90.02, 89.25, 87.14	92.53, 88.55, 87.03
ELM [35]	93.86, 95.07, 95.87	95.89, 93.15, 97.66	95.33, 94.87, 96.61
Proposed method	98.48, 99.39, 99.63	98.28, 91.07, 100	98.99, 97.54, 99.65

ANN, artificial neural network; *ELM*, extreme learning machine; *KNN*, *K*-nearest neighbor; *NB*, naïve bayes.

The performance of the proposed multiclass skin disease classification technique has been compared with other multiclass classification strategies using different classification approaches by other investigators, in Table 4.8. For the classification of the three disease classes using different classification strategy, the same amount of dataset has been used in training and testing phase, as mentioned previously. The table portrays that the proposed methodology has outperformed the other reported classification techniques for multiclass skin disease classification. The table also indicates that extreme learning machine [35] and artificial neural network [36] have achieved better performance compared to decision tree, naïve bayes [37], and *K*-nearest neighbor [33] classifier. From Table 4.9, it has been observed that the ensemble classification technique using binary classifiers for multiclass skin disease identification has achieved highly acceptable results in comparison with the other classifier models.

A comparative study of the proposed methodology with the other state-of-the-art techniques has been given in Table 4.10. Shimizu et al. [3] have reported the classification performance indices with sensitivity of 90.48%, 82.51%, 82.61%, and 80.61% for four disease classes, namely melanoma, nevi, BCC and SK, using 964 number of dermoscopic images. The authors have used color, subregion, and textural features and classified the four disease classes using flat and layered model. Oliveira et al. [7] have described a feature subset selection model for the feature diversity of shape, texture, and color-related features from dermoscopic images. The authors have reported 94.3% accuracy, 91.8% sensitivity and 96.7% specificity for the differentiation of melanoma images using ensemble model with optimum-path forest classifier and integrated with a majority voting strategy. Abuzaghleh et al. [30] have discussed fast Fourier transform and discrete cosine transform based feature extraction along with standard shape, texture, and color features for the identification of benign, atypical and melanoma diseases. In this study, authors have obtained 96.3%, 95.7% and 97.5% classification accuracy for benign, atypical and melanoma, respectively. Yu et al. [38] have extracted local convolutional features using deep residual network and fisher vector encoding technique for extraction of representative features of melanoma and nonmelanoma dermoscopic images, collected from ISBI 2016 challenge dataset. The SVM classifier has achieved 86.81% correct classification accuracy for melanoma identification. Xie et al. [39] have extracted lesions using self-generating neural network approach and quantified color, texture and border feature descriptor for further classification of melanoma and benign lesions. The reported neural network ensemble model has achieved 91.11% ACC, 83.33% SE, and 95% SP for Xanthous dataset and 94.17% ACC, 95% SE, and 93.75% SP for Caucasian race dataset. From Table 4.10, it has been observed that the proposed multiclass skin lesion classification technique has achieved improved classification performance compared to the recently published deep learning-based skin lesion identification techniques. The tabulation reflects that the proposed multiclass classification technique achieves better classification performance in comparison with the recently published work. A conclusion can be drawn from Table 4.10 that the reported methodology can work well for the representative feature extraction and multiclass classification of skin abnormalities using proposed layered structure.

4.7 Multilabel ensemble technique for multiclass skin lesion classification

This study reports a systematic approach for the feature extraction and subsequent classification of benign and malignant skin diseases of both the melanocytic and epidermal lesion categories from dermoscopic images. For this purpose, melanoma, and nevus are considered as representatives of the melanocytic skin lesion category, whereas BCC and seborrheic keratoses (SKs) are included under the epidermal lesion category. The present study explicates the extraction of spatial and the spectral features from

Table 4.10 Classification performance comparison of proposed method with state-of-art techniques.

Work	Method	Database	Number of images	Disease classes	ACC (%)	SE (%)	SP (%)
Shimizu et al. [3]	Color, subregion and textural feature extraction and layered classification model	Different Universities	964	Four classes: Melanoma, nevi, BCC, SK	–	90.48, 82.51, 82.61, 80.61	–
Oliveira et al. [7]	Color, shape and texture feature manipulation and classification using ensemble model	ISIC Archive Dataset	1104	Two classes: Melanoma and benign	94.3	91.8	96.7
Abuzaghleh et al. [30]	Color, shape, border, frequency domain, and pigment network feature extraction and classification using SVM classifier	PH2 database	200	Three classes: Benign, atypical, melanoma	96.3, 95.7, 97.5	–	–
Yu et al. [38]	Deep learning method and local descriptor encoding based feature extraction and classification using SVM classifier	ISBI 2016 challenge dataset	1279	Two classes: Melanoma and benign	86.81	–	–
Xie et al. [39]	Extraction of color, texture and border feature descriptor and classification using neural network ensemble model	Xanthous and Caucasian race dataset	600	Two classes: Melanoma and benign	91.11 94.17	83.33 95.00	95.00 93.75
Proposed method	Morphological, fractal and FRTA feature extraction and layered structure classification	ISIC Archive Dataset, PH2 database ATLAS, and, IDS	6579	Three classes: Melanoma, nevi, BCC	98.99, 97.54, 99.65	98.28, 91.07, 100	98.48, 99.39, 99.63

ACC, Accuracy; *BCC*, basal cell carcinoma; *FRTA*, fractal based regional texture analysis; *ISBI*, international symposium on biomedical imaging; *ISIC*, international skin imaging collaboration; *SK*, seborrheic keratosis; *SE*, sensitivity; *SP*, specificity; *SVM*, support vector machine.

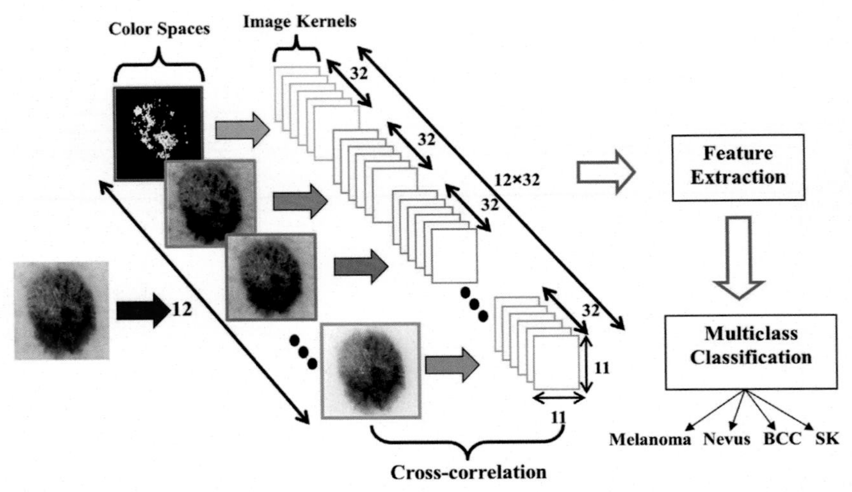

Figure 4.9 Block diagram representation of the proposed scheme. *BCC*, Basal cell carcinoma; *SK*, seborrheic keratosis.

conspicuous regions of skin lesions on the basis of similar visual impacts with the appropriate kernel patches, using the cross-correlation technique. Depending on the dermoscopic features, the kernel patches have been chosen from a set of dermoscopic images comprising all the skin disease categories selected for this work. A multilabel ensemble multiclass skin lesion classification strategy has been introduced for the segregation of malignant and benign melanocytic and epidermal skin lesions, along with their subclass classification. Prior to the classification, a cross-correlation based technique has been introduced for the extraction of spatial and spectral features from each of the images, invariant to illumination and light intensity changes. The detailed block diagram representation of the reported scheme has been depicted in Fig. 4.9.

4.7.1 Materials

To carry out the propounded work, only the dermoscopic images of various skin abnormalities corroborated by biopsy have been collected from different freely available databases across the Internet [25–28]. The entire dataset has the dermoscopic images acquired in different imaging modalities, including both easy and difficult cases. This variability of the constituents of the dataset has helped to validate the robustness of the present methodology. The entire available dataset consisting of different classes of diseases, the dermoscopic images annotated as malignant or benign and categorized as melanoma, nevus, BCC and SK have been shown in Fig. 4.10A. The entire dataset consists of 2879 number of melanoma, 3013 nevus, 796 BCC, and 513 SK dermoscopic lesions. From the entire dataset, an image subset considering all the lesion classes has been created for the selection of kernel or image patches. Considering the feature

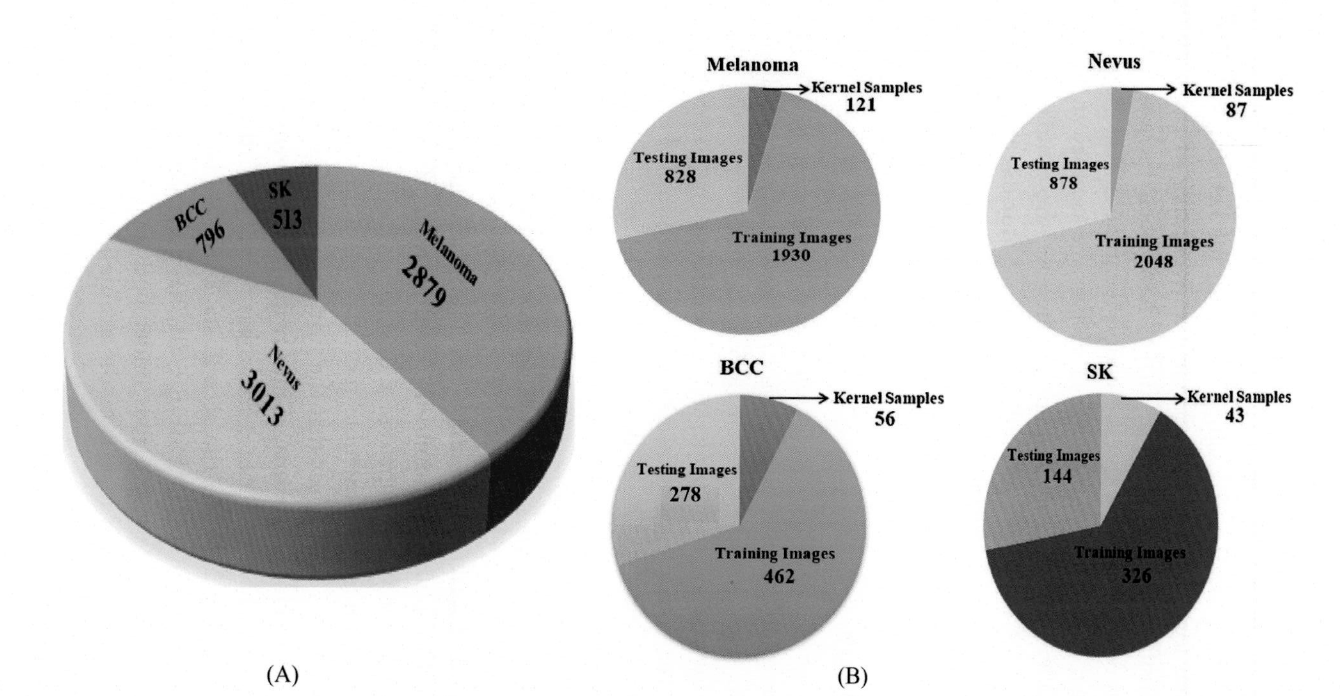

Figure 4.10 The elaborated dataset (A) original dataset, (B) dataset prepared for kernel image generation and training/testing samples of each disease classes. *BCC*, Basal cell carcinoma; *SK*, seborrheic keratosis.

variations, overall 307 number of dermoscopic images consisting of all the four disease classes have been selected for the kernel generation. After selection of the images for kernel generation, the remaining dermoscopic images have been considered for further training and testing purpose to ensure nonoverlapping between two image sets. Fig. 4.10B has depicted the detailed sample selection for kernel generation, training, and testing purposes for each class. To prevent the overfitting of the proposed model for skin lesion classification, the training images have been increased by generating "duplicate" images using traditional transformation techniques. Among the various disease classes considered, the number of dermoscopic image samples of BCC and SK are comparatively smaller than the remaining. To increase the sample size, the sample images of BCC and SK have been rotated in three and four directions, respectively, at intervals of 45°. The rotated versions of every such sample images are treated as the same image acquired in different directions or orientations.

4.7.2 Methodology

In this proposed scheme, a cross-correlation based feature extraction technique has been employed for multiclass classification of skin diseases. A multilabel ensemble multiclass classification strategy has been employed to differentiate the malignant and benign lesions of melanocytic and epidermal classes. Here, the dermoscopic images of melanoma and nevus disease classes have been considered in melanocytic lesion category, whereas BCCs and SKs in epidermal class. The objective of this multilabel ensemble multiclass classification strategy is to primarily segregate the malignant and benign lesions and further subcategorization in the consequent stage. In one-versus-all classification technique different hyperplanes are constructed, depending on the number of classes considered for classification. But this classification model is unable to segregate miss-classified images of each class and also inappropriate for subclass classification. In biomedical condition monitoring identification of misclassified images is very much important for further analysis. Considering this important aspect of biomedical applications, multilabel ensemble multiclass classification technique is introduced here. From Fig. 4.11 it is clear that the first stage classifier (CLASSIFIER I) has been employed to differentiate the images into malignant and benign classes. The segregation of malignant and benign disease classes has helped in further subcategorization with higher degree of accuracy. In the second stage of the classification model, two classifiers (CLASSIFIER II and CLASSIFIER III) have been used for the differentiation of melanocytic and epidermal lesions under both malignant and benign classes. To separate out the misclassified images or the skin lesion image other than the considered disease classes (melanoma, nevus, BCC and SK), two more classifiers (CLASSIFIER III and CLASSIFIER-IV) have been introduced in the last stage of the classification model.

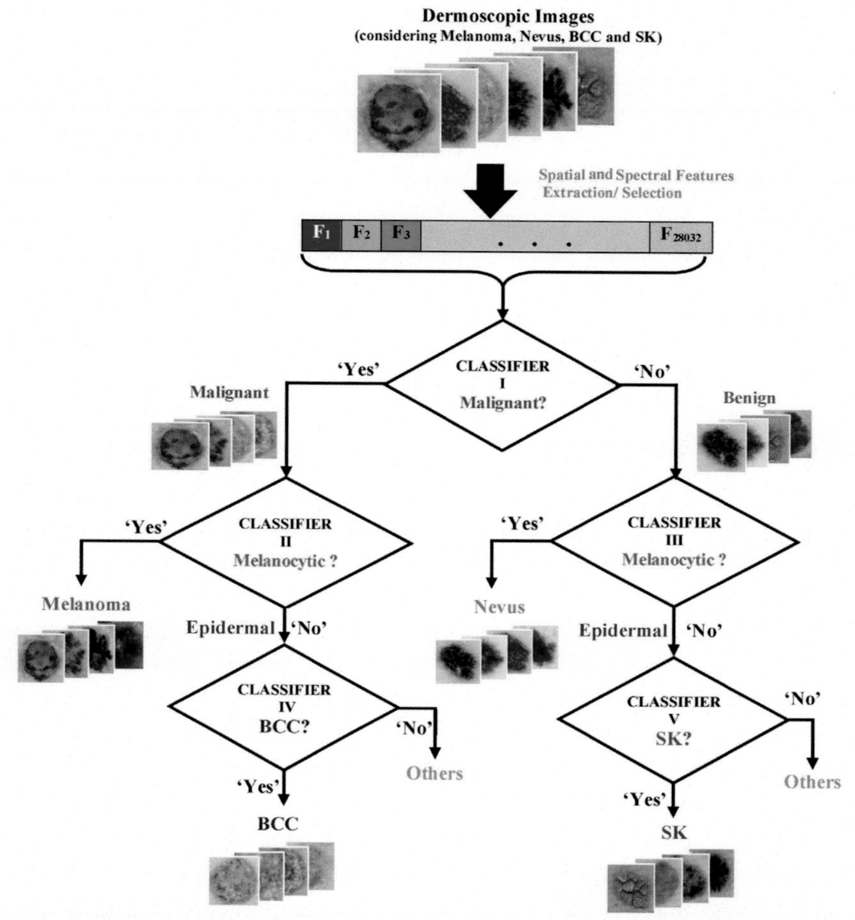

Figure 4.11 Schematic representation of the multilabel ensemble multiclass classification technique for the identification of melanoma, nevus, basal cell carcinoma (BCC), and seborrheic keratosis (SK).

For the proposed multilabel ensemble multiclass classification model, 70% images from each of the dermoscopic image classes have been considered for the construction of training samples and remaining 30% for the testing purpose. As discussed in Section 2, with the introduction of the data augmentation technique using image rotation operation, 462 and 326 number of training sample images of BCCs and SK, respectively, have increased to tackle the overfitting problem of this classification model. Prior to the classification, SVM-RFE with CBR feature selection technique has been used. As per contemporary reports [40,41], SVM emerges as an effective classifier in various biomedical engineering problems and hence has been implemented in this proposed multiclass classification problem. Here, the binary SVM classifiers have been introduced to primarily segregate benign and malignant skin lesions, and in the subsequent stages four different disease classes have been identified. To train the classification model, the tenfold cross-validation technique [10] has been used. During the training phase, 1930 number of melanoma, 2048 number of nevus, 2072 BCC and 1880 SK images have been divided into 10-folds, where each of the images have been considered in training and validation set. After training each of the binary SVM classifier, a total of 2072 dermoscopic images of four different classes has been identified in the testing phase. The entire feature extraction and classification algorithm has been implemented using MATLAB version 2017b deployed in a computer with 64-bit Operating System, x64-based processor having 16 GB RAM, Intel Core i5−7400@3.00 GHz processor.

4.7.3 Results and discussions

The cross-correlation operation of the dermoscopic image with several kernels in different color channels yields significant subregions of the input image, having similar dermoscopic feature as the selected kernel patches. Consulting with an expert dermatologist, the specific dermoscopic finding have been identified from 307 number of dermoscopic images, comprised of melanoma, nevus, BCC and SK disease classes, for the selection of the kernels. The regional spatial and spectral features have described the presence of the similar dermatological findings in the corresponding dermoscopic images. To ensure the nonoverlapping image sets, the overall number of dermoscopic images (6838), excluding the sample images considered for kernel generation, have been studied for feature extraction and further classification of the diseases. To reduce the overfitting problem of the classification model, during the training phase, the data augmentation techniques have been used. As mentioned in previous sections, 70% images of the dataset have been used as the training samples for the multilabel ensemble multiclass classification model.

In the reported multilabel ensemble model, the binary SVM classifiers have been used for each of the classification stages. To classify the dermoscopic images from linearly nonseparable feature set, the nonlinear SVM kernels have been chosen. For the proper selection of kernel function and the corresponding nonlinear kernel hyperparameters, investigations have been carried with two hyperparameter optimization algorithms selected for the optimization of penalty parameter of soft margin cost function and the scale parameter of the kernel. The performances of these algorithms, namely grid-search-based methods and Bayesian optimization techniques [41], are presented in Table 4.11.

It can also be easily seen that in this study, the RBF and polynomial kernels of different orders have been considered for the SVM classifier. From the comparative performance analysis, it has been found that the higher classification performance indices

Table 4.11 Comparative study on different kernel parameter optimization techniques for nonlinear kernels of binary SVM classifiers.

Kernel	Kernel parameters optimizer					
	Grid search method			Bayesian optimization		
	% SE	% SP	% ACC	% SE	% SP	% ACC
RBF	98.97	91.74	98.21	99.01	95.35	98.79
Polynomial (2nd order)	98.39	90.38	97.39	98.63	91.57	97.76
Polynomial (3rd order)	98.51	91.34	97.65	98.74	92.74	98.01

ACC, Accuracy; *SE*, sensitivity; *SP*, specificity.

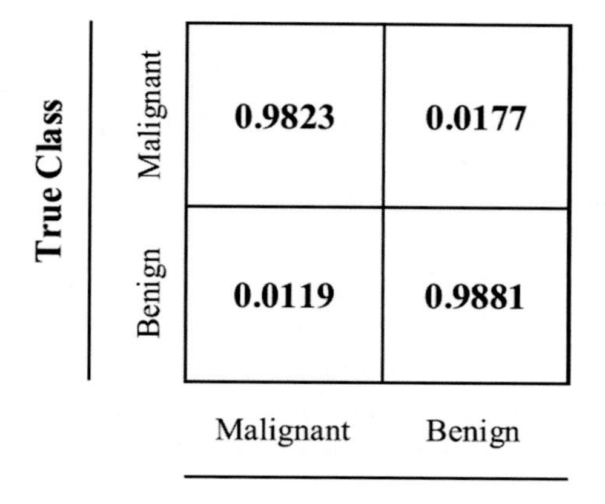

Figure 4.12 Confusion matrix for the classification of benign and malignant skin lesions.

have been obtained for the SVM classifier with RBF kernel for both the optimization techniques. The Bayesian optimization method has outperformed the grid search algorithm for hyperparameter optimization of RBF kernel with 99.01% SE, 95.35% SP and 98.79% ACC for the identification of four disease classes. Multilabel ensemble multiclass classification model has been implemented using binary SVM classifiers with RBF kernel. To train the classification model, properly labeled 7930 number of dermoscopic images have been used in tenfold cross-validation technique. The first label of the classification model has separated the benign and malignant lesions from the entire dataset. In Fig. 4.12 the confusion matrix of the first stage classifier (CLASSIFIER I) has revealed that the malignant and benign lesions have been identified with higher degree of sensitivity, The high sensitivities of 98.23% and 98.81% for the identification of melanoma and benign lesions, respectively, have ensured the better classification performance in the subsequent stages. The detailed confusion matrix of this classification strategy has been shown in Fig. 4.13. In the confusion matrix, the

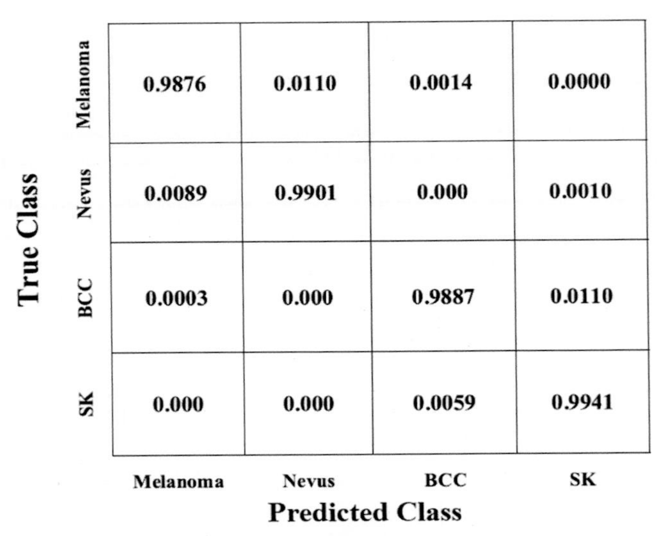

Figure 4.13 Confusion matrix for four class classification of melanoma, nevus, basal cell carcinoma (BCC), and seborrheic keratosis (SK) dermoscopic images.

four compartments in each of the true and predicted classes have been used to represent the corresponding disease classes as melanoma, nevus, BCC and SK under the melanocytic and epidermal lesion category. The confusion matrix has revealed that the melanoma and nevus of melanocytic lesions type have been classified with the sensitivity of 98.76% and 99.01%, respectively. Similarly, 98.87% and 99.41% sensitivities have been achieved for the identification of BCC and SK lesions. The subcategorization of malignant and benign lesions with higher classification indices has justified the use of multilabel ensemble classification strategy as an efficient multiclass classification technique.

The classification performance indices significantly describe the correct classification rate of the malignant melanoma, nevus, BCC, and SKs. The classification performance of the first stage of the two stage classification model for the identification of malignant and benign skin lesions have been compared with the dermatological expert's decisions. Each of the dermoscopic images, labeled by dermatologists, has been shown to the expert to identify whether a lesion is malignant or benign.

For the identification of malignant and benign lesions, the comparative representation of the classification performance indices has been shown in Fig. 4.14A. The malignant and benign lesions have been classified with 98.80% SE, compared to 86.01% by the expert. The algorithm has attained 98.83% SP at the first stage of the classification model. The method also attains higher correct classification accuracy of 98.52% by the proposed technique, compared to those obtained by expert as 85.43%. Higher classification performance indices in the first stage of the binary classification ensure significant improvement in the further subclass classification of the dermoscopic

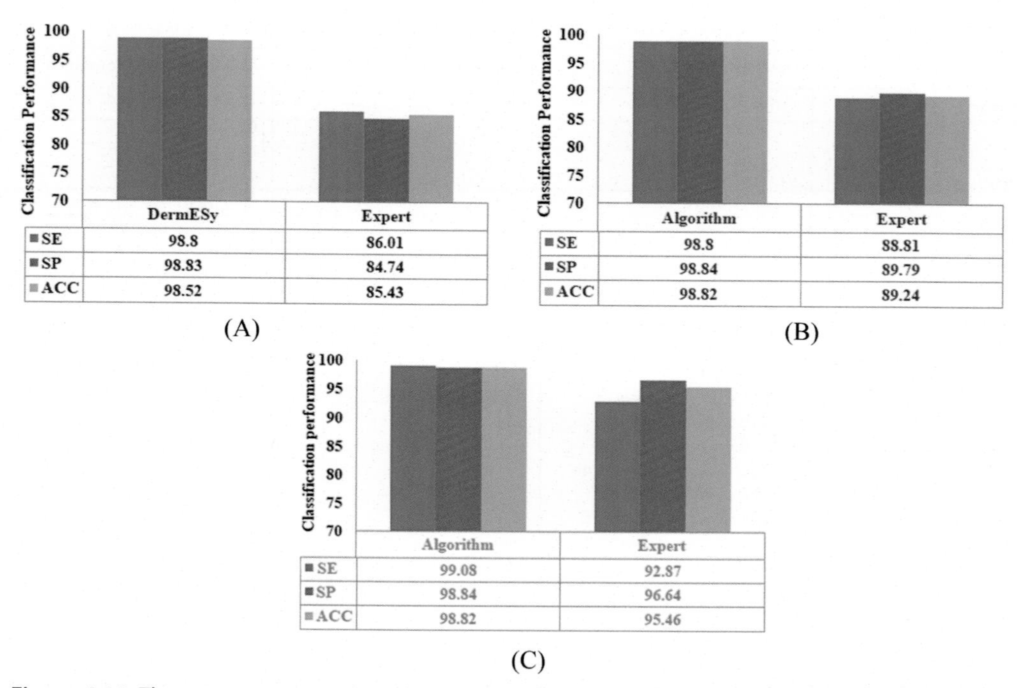

Figure 4.14 The comparative analysis on classification performance indices among proposed algorithm and dermatological experts' (A) malignant versus benign; (B) malignant melanocytic versus malignant epidermal; (C) benign melanocytic versus Benign epidermal; where sensitivity, specificity and accuracy have been specified as SN, SP and ACC, respectively.

images. It is also found, that the dermoscopic images of melanocytic and epidermal lesion category, have also been classified precisely.

In this study, the malignant melanoma and BCC have been considered as the representatives of malignant melanocytic and malignant epidermal lesions, respectively, whereas atypical nevus and seborrheic keratoses have been assigned to the benign melanocytic and epidermal lesion category. The performance of the subclass classification of this reported classification model has been compared with the performance of an expert dermatologist. The complex and similar visual appearance of the skin lesions make it difficult for the dermatological expert to differentiate by visual impression only. To differentiate malignant melanocytic and epidermal lesions, the multiclass classification model has achieved 98.80% SE, 98.84% SP, and 98.82% ACC. The expert dermatologist has identified the malignant melanocytic and epidermal lesions from visual inspection, with a correct classification accuracy of 89.24%, as shown in Fig. 4.14B. The expert has also been assigned the task of identifying the benign melanocytic and epidermal lesions from the same image set. To the expert, the differentiation of these two categories has appeared to be more easily distinguishable due to their

Table 4.12 Performance comparison of the proposed method with state-of-the-art techniques.

Work	Database	Disease classes	Classification performance
Oliveira et al. [7]	ISIC: dataset	Malignant and benign	Accuracy—94.3% Sensitivity—91.8% Specificity—96.7%
Kasmi and Mokrani [29]	EDRA, Interactive Atlas of Dermoscopy	Malignant and benign	Sensitivity—91.25% Specificity—95.83%
Rastgoo et al. [4]	Vienna General Hospital	Melanoma and dysplastic nevi	Sensitivity—98% Specificity—70%
Shimizu et al. [3]	Keio University Hospital, University of Naples and Graz, Tokyo Women's Medical University	Melanoma, nevus, BCC, and SK	Sensitivities—90.48%, 82.51%, 82.61% and 80.61%
Gonzàlez-Dìaz [9]	2017 ISBI Challenge database	Melanoma, nevus, and SK	Sensitivity—95%
Esteva et al. [28]	ISIC Dermoscopic Archive, the Edinburgh Dermofit Library and data from the Stanford Hospital	Benign, malignant, and nonneoplastic	Sensitivity—72.1%
Present study	International Dermoscopic Society database, Dermoscopic Atlas database, ISIC: challenge 2017, and PH2 database	Melanoma, nevus, BCC, and SK	Sensitivities—98.76%, 99.01%, 98.87% and 99.41%

BCC, Basal cell carcinoma; *ISBI*, international symposium on biomedical imaging; *ISIC*, international skin imaging collaboration; *SK*, seborrheic keratosis.

distinctive visual appearance. The same has been revealed in Fig. 4.14C. Higher classification performances have been reported by the dermatological expert, with 92.87% sensitivity, 96.64% specificity and 95.46% accuracy. The classification performance for benign melanocytic and epidermal lesions has been improved by the proposed algorithm, with 99.08% sensitivity, 98.84% specificity and 98.82% correct classification accuracy.

In Table 4.12 the proposed methodology for the classification of skin abnormalities has been compared with some state-of-the-art techniques. The input feature manipulation processes based ensemble model, proposed by Oliveira et al. [7] has achieved 94.36% accuracy, 91.8% sensitivity and 96.7% specificity for the classification of malignant and benign skin lesions. Kasmi and Mokrani [29] have implemented the algorithms to extract the characteristics of ABCD attributes to differentiate malignant melanoma and benign skin lesions with 91.25% SE and 95.83% SP. Rastgoo et al. [4] have differentiated melanoma and dysplastic nevi with the sensitivity of 98% and specificity of 70% using the global and local feature extraction approaches. In literature, side by side with binary classification, different multiclass skin lesion classification strategies have also been proposed. Shimizu et al. [3] have achieved 90.48%, 82.51%, 82.61%, and 80.61% detection rate for melanomas, nevi, BCC and SK, respectively, using the layered model. Gonzàlez-Dìaz [9] has reported a system for the identification of melanoma, nevus, and SKs with the sensitivity of 95%, using convolutional neural network approach. Esteva et al. [42] have achieved three-way accuracy of 72.1% for the classification of benign, malignant and nonneoplastic lesions. In the study presented here, the classification of malignant and benign lesions of both melanocytic and epidermal category has been attained with significantly reduced misclassification rate. This is amply indicated by the corresponding sensitivity values of 98.76%, 99.01%, 98.87%, and 99.41%.

4.8 Conclusion

The primary objective of this section is to identify and monitor the skin abnormalities from dermoscopic images using digital signal processing tools. As discussed in the previous chapter, employing different efficient feature extraction tools, a wide range of morphological, texture, and color-related features have been extracted. From this representative information of various skin diseases, classification models have been introduced to differentiate various skin abnormalities. In this chapter, classification techniques for the differentiation of different skin abnormalities have been introduced. To identify the diseases with lesser number of features, the most efficient features have been selected by employing feature selection algorithms. In this chapter, SVM-RFE technique is used. To obtain reduced set of effective features from large number of feature set, automatic CBR algorithm is introduced. Employing SVM-RFE with CBR technique, melanoma, and benign nevus diseases have been differentiated with 97.63% sensitivity, 100% specificity and 98.28% accuracy. For the identification of three or more skin disease classes, multiclass classification strategy is reported here. Employing SVM-RFE feature selection technique, an SVM-based multiclass classification model is developed for the identification of melanoma, nevus, and BCC lesions from the dermoscopic images. This classification model is also able to segregate the miss-classified images for further monitoring of diseases. Using this classification technique, three disease classes have been identified with a highly acceptable classification accuracy of 98.99%, 97.54% and 99.65% for melanoma, dysplastic nevi and BCC, respectively. Multilabel ensemble multiclass skin lesion classification technique is introduced to classify melanoma, nevus, BCC, and SK diseases from hystopathologically confirmed dermoscopic images. To develop this classification model, five binary SVM classifiers associated with feature selection technique are used. In this classification model, the four disease classes have been differentiated by stage wise segregation of melanocytic and epidermal lesions with subclass categorization. It has been possible to identify both malignant and benign lesions of melanoma, nevus, BCC, and SK disease classes, with high sensitivities and reveals the potential of the technique to assist the dermatologists and expert clinicians in further decision making.

References

[1] R. Marks, An overview of skin cancers: incidence and causation, Cancer Suppl. 75 (S2) (1995) 607–612.

[2] I. Maglogiannis, K. Delibasis, Enhancing classification accuracy utilizing globules and dots features in digital dermoscopy, Comput. Method Prog. Biomed. 118 (2015) 124–133.

[3] K. Shimizu, H. Iyatomi, M.E. Celebi, K. Norton, M. Tanaka, Four-class classification of skin lesions with task decomposition strategy, IEEE Trans. Biomed. Eng. 62 (1) (2015) 274–283.

[4] M. Rastgoo, R. Garcia, O. Morel, F. Marzani, Automatic differentiation of melanoma from dysplastic nevi, Comput. Med. Imaging Graph. 43 (2015) 44–52.

[5] C. Barata, M. Ruela, M. Francisco, T. Mendonc, J.S. Marques, Two systems for the detection of melanomas in dermoscopic images using texture and color features, IEEE Syst. J. 8 (3) (2014) 965–979.

[6] A. Sáez, J. Sánchez-Monedero, P.A. Gutiérrez, C. Hervás-Martínez, Machine learning methods for binary and multiclass classification of melanoma thickness from dermoscopic images, IEEE Trans. Med. Imaging 35 (4) (2016) 1036–1045.

[7] R.B. Oliveira, A.S. Pereira, J.M.R.S. Tavares, Skin lesion computational diagnosis of dermoscopic images: ensemble models based on input feature manipulation, Comput. Method Prog. Biomed. 149 (2017) 43–53.

[8] D. Ravì, C. Wong, F. Deligianni, M. Berthelot, J. Andreu-Perez, B. Lo, et al., Deep learning for health informatics, IEEE J. Biomed. Health Inform 21 (1) (2017) 4–21.

[9] I. González-Díaz, DermaKNet: incorporating the knowledge of dermatologists to Convolutional Neural Networks for skin lesion diagnosis, IEEE J. Biomed. Health Inform. 23 (2) (2019) 547–559 (2018).

[10] B. Harangi, Skin lesion classification with ensembles of deep convolutional neural networks, J. Biomed. Inform. 86 (2018) 25–32.

[11] J. Kawahara, G. Hamarneh, Fully convolutional neural networks to detect clinical dermoscopic features, IEEE J. Biomed. Health Inform. 23 (2) (2019) 578–585.

[12] Y. Yuan, Y. Lo, Improving dermoscopic image segmentation with enhanced convolutional–deconvolutional networks, IEEE J. Biomed. Health Inform. 23 (2) (2019) 519–526.

[13] J. Kawahara, S. Daneshvar, G. Argenziano, G. Hamarneh, 7-Point checklist and skin lesion classification using multi-task multi-modal neural nets, IEEE J. Biomed. Health Inform. (2018). Available from: https://doi.org/10.1109/JBHI.2018.2824327.

[14] J.G. Carbonell, R.S. Michalski, T.M. Mitchell, An overview of machine learning, Machine Learning, Tioga Publishing Company, 1983, pp. 3–23.

[15] E. Alpaydin, Introduction to Machine Learning, The MIT Press, Cambridge, Massachusetts London, England, 2004.

[16] F. Camastra, A. Vinciarelli, Machine Learning for Audio, Image and Video Analysis: Theory and Analysis, Springer, 2008.

[17] I. Guyon, S. Gunn, M. Nikravesh, L.A. Zadeh, Feature Extraction Foundations and Applications, Springer, 2006.

[18] V. Vapnik, The Nature of Statistical Learning Theory, Springer- Verlag, New York, 1995.

[19] I. Guyon, J. Wetson, S. Barnhill, V. Vapnik, Gene selection for cancer classification using support vector machines, Mech. Learn. 46 (1–3) (2002) 389–422.

[20] M. Li, W. Chen, T. Zhang, Automatic epileptic EEG detection using DT–CWT-based non-linear features, Biomed. Signal Proc. Control. 34 (2017) 114–125.

[21] Y. Tang, Y. Zhang, Z. Huang, Development of two-stage SVMRFE gene selection strategy for microarray expression data analysis, IEEE/ACM Trans. Comput. Bio. Bioinform. 4 (3) (2007) 365–381.

[22] B.L. Koley, D. Dey, On-line detection of apnea/hypopnea events using SpO2 signal: a rule-based approach employing binary classifier models, IEEE J. Biomed. Health Inform. 18 (1) (2014) 231–239.

[23] K. Yan, D. Zhang, Feature selection and analysis on correlated gas sensor data with recursive feature estimation, Sens. Actuators B 212 (2015) 353–363.

[24] L. Tolosi, T. Lengauer, Classification with correlated features: unreliability of feature ranking and solutions, Bioinformatics 27 (2011) 1986–1994.

[25] International Dermoscopy Society. <http://www.dermoscopy-ids.org>.
[26] Dermoscopy Atlas. <http://www.deroscopyatlas.com>.
[27] T. Mendonça, P.M. Ferreira, J. Marques, A.R.S. Marcal, J. Rozeira, PH2—a dermoscopic image database for research and benchmarking,35th International Conference of the IEEE Engineering in Medicine and Biology Society, July 3–7, 2013, Osaka, Japan.
[28] D. Gutman, et al., Skin lesion analysis toward melanoma detection: Achallenge at the international symposium on biomedical imaging (ISBI) 2016, hosted by the international skin imaging collaboration (ISIC), 2016 [Online] Available: <https://arxiv.org/abs/1605.01397>.
[29] R. Kasmi, K. Mokrani, Classification of malignant melanoma and benign skin lesions: implementation of automatic ABCD rule, IET Image Process. 10 (6) (2016) 448–455.
[30] O. Abuzaghleh, B.D. Barkana, M. Faezipour, Non-invasive realtime automated skin lesion analysis system for melanoma early detection and prevention, IEEE J. Trans. Eng. Health Med. 3 (2015).
[31] R. Garnavi, M. Aldeen, J. Bailey, Computer-aided diagnosis of melanoma using border and wavelet-based texture analysis, IEEE Trans. Inform. Technol. Biomed. 16 (6) (2012) 1239–1252.
[32] U.R. Acharya, S.V. Sree, P.C. Alvin Ang, R. Yanti, J.S. Suri, Application of non-linear and wavelet based features for the automated identification of epileptic EEG signals, Int. J. Neural Syst. 22 (2) (2012).
[33] U.R. Acharya, Y. Hagiwara, S.N. Deshpande, S. Suren, J.E.W. Koh, S.L. Oh, et al., Characterization of focal EEG signals: a review, Future Gen. Comput. Syst. 91 (2019) 290–299.
[34] M. Sharma, D. Deb, U.R. Acharya, A novel three-band orthogonal wavelet filter bank method for an automated identification of alcoholic EEG signals, Appl. Intell. 48 (5) (2018) 1368–1378.
[35] G. Bin Huang, Q.Y. Zhu, C.K. Siew, Extreme learning machine: theory and applications, Neurocomputing 70 (1–3) (2006) 489–501.
[36] C.M. Bishop, Pattern Recognition and Machine Learning, Springer, 2006. ISBN-10:0-387-31073-8.
[37] L. Breiman, J.H. Friedman, R.A. Olshen, C.J. Stone, Classification and Regression Trees, Chapman & Hall, Boca Raton, FL, 1984.
[38] Z. Yu, X. Jiang, F. Zhou, J. Qin, D. Ni, S. Chen, et al., Melanoma recognition in dermoscopy images via aggregated deep convolutional features, IEEE Trans. Biomed. Eng. 66 (4) (2019) 1006–1016.
[39] F. Xie, H. Fan, Y. Li, Z. Jiang, R. Meng, A. Bovik, Melanoma classification on dermoscopy images using a neural network ensemble model, IEEE Trans. Med. Imaging 36 (3) (2017) 849–858.
[40] S. Lahamiri, An accurate system to distinguish between normal and abnormal electroencephalogram records with epileptic seizure free intervals, Biomed. Signal Process. Control. 40 (2018) 312–317.
[41] S. Lahamiri, A. Shmuel, Detection of Parkinson's disease based on voice patterns ranking and optimized support vector machine, Biomed. Signal Process. Control. 49 (2019) 427–433.
[42] A. Esteva, B. Kuprel, R.A. Novoa, J. Ko, S.M. Swetter, H.M. Balau, et al., Dermatologist-level classification of skin cancer with deep neural network, Nature 542 (2017) 115–118.

CHAPTER 5

Development of expert system for skin disease identification

5.1 Introduction

The early and accurate diagnosis of melanoma is important for further treatment and prevention of the disease. The complex structure, multiplicity of color and similarity in visual appearance make it difficult to differentiate malignant melanocytic lesions from other skin abnormalities [1]. Dermatologists use epiluminescence microscopy or dermoscopy, a noninvasive and noncontact imaging technique, for the in-depth visualization of the pigmented area. The introduction of dermoscopy with detailed visualization of the skin lesion has increased the diagnostic accuracy by 5%–30% [2]. Incorporating this screening tool, the dermatologists used clinical algorithms such as ABCD rule [3], and seven-point checklist [4,5] for the diagnosis of the skin diseases. The computer aided diagnostic system has come up with great possibilities in accurate diagnosis with quantification of the dermoscopic findings. The recent trends in the development of computer aided diagnosis of skin diseases have assisted the clinicians and experts in investigating an out-sized number of patients with higher diagnostic accuracy. The ABCD rule of dermoscopy has been developed to express the dermoscopic findings quantitatively for the diagnosis of the lesion under investigation. The ABCD rule of dermoscopy has incorporated the clinical criterion of the skin lesions such as asymmetry, border irregularity, color, and differential structures. Combining these dermoscopic criteria, the total dermoscopic score (TDS) has been evaluated for the grading of the lesions as benign, suspicious or malignant. Computer aided diagnostic system has evaluated the ABCD criteria and improved the diagnostic performance not only for the experts but also for clinicians with limited experience in dermoscopy. Kasmi and Mokrani [6] has implemented the ABCD rule for the classification of malignant melanoma and benign skin lesions. For the estimation of the TDS, authors have used the shape, brightness, and color asymmetry along with structural features for the improvement of the classification performance. Literature suggests different methodologies for the automatic diagnosis of skin lesions, implementing hand-crafted features [7–11]. The recent development in deep learning techniques have been used for skin disease identification by computer vision, with impressive performance [12–14]. González Díaz has introduced a convolutional neural network-based approach, incorporating the knowledge of dermatologists for the identification of the skin lesions [15].

Recent Trends in Computer-aided Diagnostic Systems for Skin Diseases
DOI: https://doi.org/10.1016/B978-0-323-91211-2.00003-2

The proposed system has lesion segmentation and dermoscopic structure segmentation blocks and identified melanoma and seborrheic keratosis (SK) diseases with area under the curve of 87.3 and 96.2, respectively. In this chapter, a dermatological expert system (DermESy) for skin disease identification has been propounded. This expert system tool aims at assisting clinician and experts to provide a reliable, accurate and quick diagnosis along with monitoring of the skin lesions. To develop an indigenous integrated system with convenient visual monitoring of the dermoscopic findings, the following computing aids have been introduced in this study:

- Simple and efficient algorithms have been implemented to score the ABCD attributes.
- The estimation of TDS is reframed by improvised ABCD attributes. In addition to asymmetry in shape, brightness, and color variations are also incorporated to reform the "A" score.
- For the determination of "C" score, a novel color information extraction module (CIEM) incorporating expert's knowledge has been introduced to detect the significant dermoscopic colors. For the identification of significant color regions, the colors of the corresponding expert's annotated regions from the same dataset are considered as reference.
- For pigment network and structureless area detection, mathematical morphology aided filtering technique is introduced. The irregular or localized distribution of the structure in the lesion area is more indicative toward melanoma than globally distributed structures. Considering this dermoscopic criterion, the "D" score has been evaluated by not only considering the presence of the differential structures but also their spatially localized information.
- An explanatory subsystem is introduced to give the interpretations of the TDS so as to enable the analysis of the dermoscopic findings, and thus assists the expert to cross-verify the diagnosis by the DermESy.
- All the histopathologically confirmed dermoscopic images have been given to the dermatologist to segregate the malignant, benign and suspicious lesions. Dermatologist's opinion has been compared with the performance of the DermESy for the differentiation of malignant and benign lesions. The dermoscopic findings as extracted from DermESy are correlated with the dermatologist's assessment and a considerable improvement in diagnosis is successfully achieved, which satisfies the primary objective of this study.

5.2 Knowledge base for skin disease diagnosis

Expert system is an integrated computer aided decision-making system where expert knowledge, experience and skill are implemented using a rule-base for drawing proper

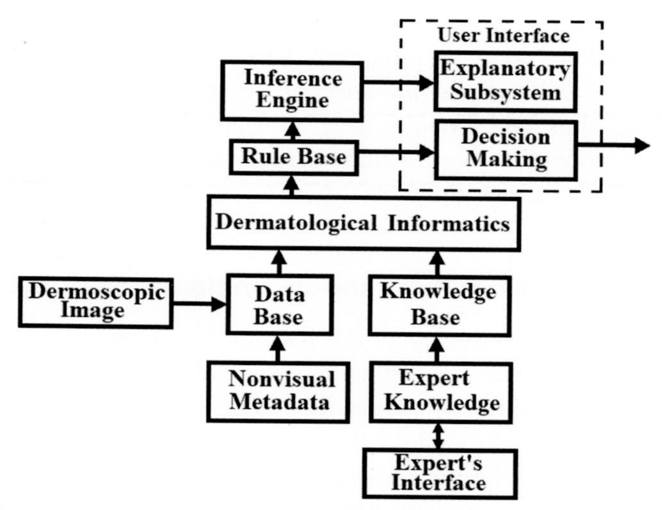

Figure 5.1 Block diagram of the dermatological expert system.

inferences. A diagnostic expert system emulates the expertise of an experienced doctor, having specific domain knowledge, essentially by the following major components:

- knowledge base and rule base
- inference engine
- user interface

The block diagram of a DermESy has been shown in Fig. 5.1. The knowledge base of the DermESy contains the experts' understanding based on clinical "naked eye" observation of the skin abnormalities. It has replicated the dermatologist's knowledge to formulate the ABCD rule for the characterization of skin lesions.

5.2.1 ABCD rule for skin disease identification

The ABCD algorithm is developed to quantitatively address the crucial dermoscopic criteria such as asymmetry (A), border irregularity (B), color (C), and differential structures (D).

- *Asymmetry*: The structural variation along the entire lesion is considered to differentiate the benign and the malignant lesions clinically. The criterion encompasses both contour asymmetry and the disproportionate distribution of dermoscopic color, luminance, and structures along the lesion area.
- *Border*: To assess whether there is an abrupt cut off of pigment pattern at the periphery of the lesion, or a gradual indistinct cutoff, the border irregularity has been estimated.
- *Color*: The presence of one or more colors or an uneven distribution of color is taken into account by the dermatologist for the early-stage diagnosis of the lesion.

- *Dermoscopic structures*: For the characterization of skin lesion, presence of differential structures is considered to be an important dermoscopic finding. To determine the dermoscopic structural features, presence of structureless area, pigment network, branched streaks, dots, and globules have been given the major consideration. To estimate the dermoscopic score for differential structures, a maximum of five score is allotted, one score to each structure.

5.3 Dermatological informatics module implementing ABCD rule

The dermatological informatics module (DIM) contains the significant information regarding the skin lesion characteristics. This block is dedicated to extract the essential dermoscopic findings for logical construction of rule base. The main pipeline of the dermoscopic ABCD clinical feature extraction is depicted in Fig. 5.2. For each clinical case, the dermoscopic image *I* is preprocessed for the removal of noise and hair artifacts. The preprocessed image is considered by the lesion segmentation module (LSM) to segregate the region of interest (ROI). From the segmented image, shape asymmetry of the lesion has been evaluated to compute the asymmetry index ("A" attribute of ABCD rule). The segmented image has been fed to the border detection module (BDM) for outlining the single pixel border of the lesion area. The border detected image is used to estimate the border irregularity ("B" attribute of ABCD rule) of the lesion area. Original color dermoscopic image has been masked with the segmented image to separate out the clinically significant dermoscopic color regions from the lesion area using the CIEM. CIEM subsystem helps to determine the color asymmetry of the lesion ("A" attribute of ABCD rule) and also to find the dermoscopic colors present in the lesion ("C" attribute of ABCD rule). Segmented image masked with the original grayscale image is used to estimate the brightness asymmetry ("A" attribute of ABCD rule) present in the lesion area. The same

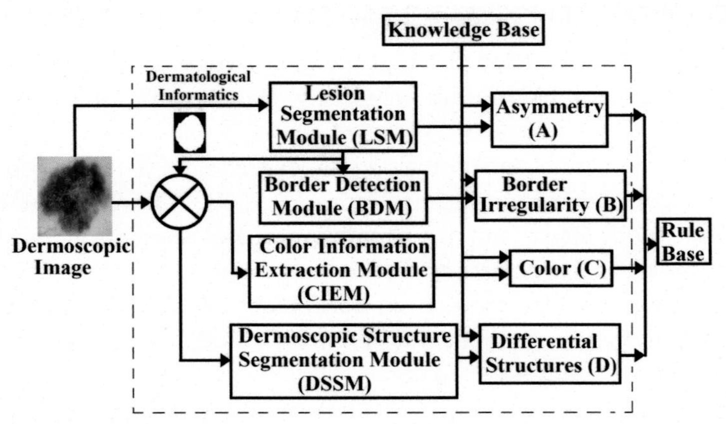

Figure 5.2 Detailed block diagram of the dermatological informatics module.

grayscale image has been used as an input to the dermoscopic structures segmentation module (DSSM) to extract differential structures from the skin lesion. The dermoscopic structures segmented using DSSM subsystem has been used to compute the "D" score of the ABCD rule. For the estimation of the "D" score, spatial information of the differential structures has been considered to develop an improvised ABCD rule of dermoscopy. Description of all the modules considered for the excavation of ABCD dermoscopic features have been given in detail in the subsequent sections.

5.3.1 Image preprocessing

In the preprocessing stage, a median filter is used to eliminate the noise due to uneven illumination. Dermoscopic images of skin lesions from various anatomic sites are contaminated with thick or thin hairs. To remove these hair artifacts, the morphological bottom-hat filter [16] followed by the inward interpolation technique has been implemented, as discussed in Chapter 2, Preprocessing and segmentation of skin lesion images. Presence of hair structures through the lesion border region extends the lesion area and introduces unwanted oversegmentation of the region. Elimination of hair artifacts helps to obtain appropriate segmented region for further analysis and estimation of dermatological properties of the lesion.

5.3.2 Lesion segmentation module

After preprocessing, segmentation of lesion area is an essential step for skin disease identification and monitoring of further spreading of the disease. In DIM, the LSM is implemented to segregate the affected area from the dermoscopic image. Dermatologists mark the lesion area manually to estimate the amount of pigmentation compared to that in the normal skin area. Proper segmentation of the lesion helps to determine the morphological properties of the pigmented area and to evaluate the asymmetry index of the ROI. Mathematical morphology aided skin lesion segmentation algorithm has been developed in the LSM subsystem. The fundamental element of mathematical morphology is the structuring element (SE) [16]. Selection of appropriate SE is important for the development of algorithms using morphological operations. The shape of the SE should be selected according to the morphological properties of the object under consideration. In this study, most of the skin lesions are circular in nature or closer approximation of a circle. Therefore, to segmentize the skin lesions using morphological operations, circular SE or a circular kernel comparatively smaller in size has been considered. For the development of morphological filters and segmentation algorithm, same form of circular SE has been considered for all the dermoscopic images of the entire dataset. For the development of mathematical morphology aided segmentation algorithm, the size of the circular SE has been chosen with a diameter of eight pixels. Grayscale morphological closing operation has been performed on the original grayscale image to remove smaller objects compared

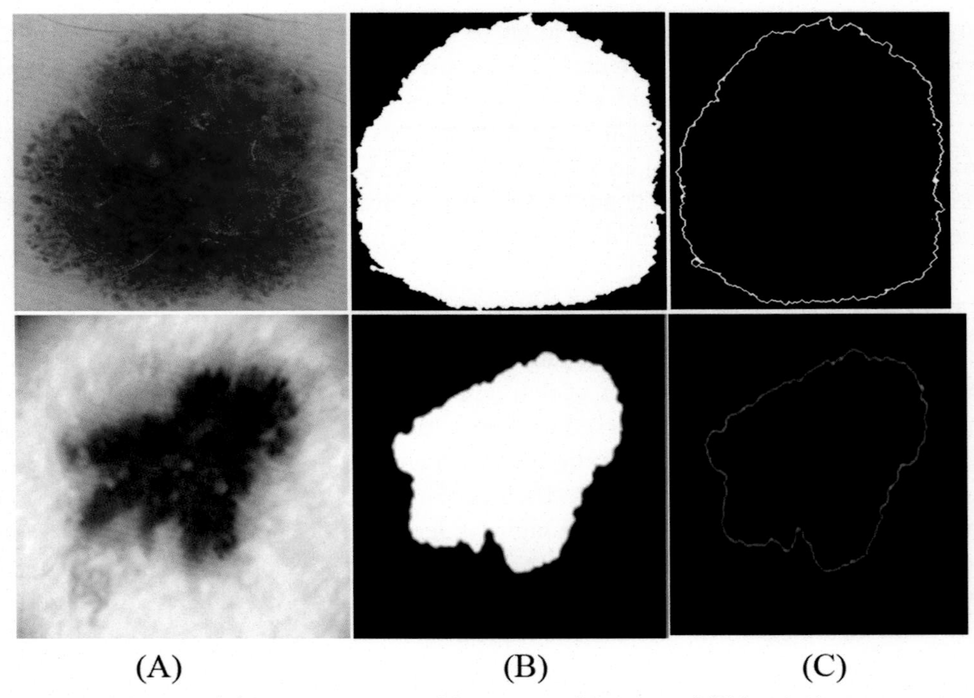

Figure 5.3 (A) Original dermoscopic image, (B) segmented image, and (C) border image.

to the size of the SE. Subtracting the complement of the image from the closed image results a sharp contrast between the lesion and surrounding background. The maximum interclass variance of the image is measured to estimate a threshold value for the segregation of the ROI. In Fig. 5.3, segmented regions of sample dermoscopic images of malignant and benign category have been shown. From the figure, it has been observed that even having significant variations in morphological properties, the reported segmentation algorithm has effectively segmented the appropriate ROI.

5.3.3 Border detection module

The segmented image (I_s) is considered in the subsequent border detection module to obtain a single pixel border of the lesion area. For the determination of single pixel border of the segmented region, two-pixel diameter has been considered for the circular SE. Selection of larger SE thickens the border region of the segmented image and the proper estimation of the irregularity is not possible [16]. The border detection algorithm has been comprehensively discussed in Chapter 2, Preprocessing and Segmentation of Skin Lesion Images. The structural variations obtained from the resultant border image is used to estimate its irregularity. The corresponding border regions of the segmented lesions have been shown in Fig. 5.3C.

5.3.4 Color information extraction module

As explained in the previous section, the dermatologist has explored the occurrence of six different colors (white, red, light brown, dark brown, blue–gray, and black) for the diagnosis of the skin lesion. This color information has been extracted from a set of sample images, in consultation with an expert dermatologist. A set of sample dermoscopic images of different classes are given to the experienced dermatologist to mark the above mentioned significant color regions. In Fig. 5.4, the ground truth (GT) images, with light brown, dark brown, red, and black regions identified and marked, have been shown. The intensity values of each color channels for the respective color have been considered as reference information for the development of the color information extraction algorithm.

The input dermoscopic image masked with the corresponding segmented image (I, I_s) has been considered to develop the color information extraction algorithm. To subdivide the skin lesion area with similar color regions, superpixels are generated by applying simple linear iterative clustering (*SLIC*) algorithm [17]. Here, the desired size of the superpixel is chosen as the input parameter in contrast to the number of superpixels as considered in *SLIC* algorithm. For an image with total P pixels, the number of superpixels, each with N pixels, is P/N. The histogram (*hist*) of an image describes the frequency of occurrence of the intensity present in that image. Therefore, in histogram, the intensity value with maximum number of bins determines the dominant color information of each superpixel region. The distance (Dc) between the prevailing color (C_{RGB}) in a superpixel region and the corresponding reference has been estimated. If this measured distance is less than the just noticeable distance [6], the entire superpixel region will be replaced by the dominant color. In Algorithm 5.1, the color information extraction technique has been described. For two dermoscopic sample images, their corresponding superpixel regions and extracted color regions have been shown in Fig. 5.5.

Algorithm 5.1:
Color information extraction

Input: Dermoscopic image I in RGB color plane; R, G, B values of six reference colors, W (white), R (red), LB (light brown), DB (dark brown), BG (blue–gray), B (black); segmented image I_s; superpixels containing N number of pixels.

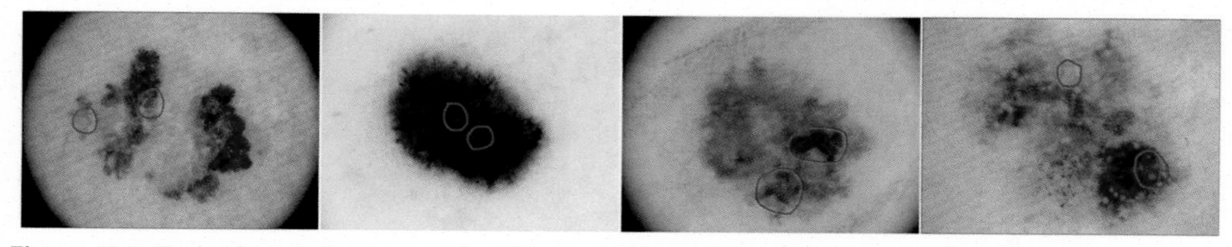

Figure 5.4 Ground truth images with different color regions (light brown, dark brown, red, black) marked by the expert dermatologist.

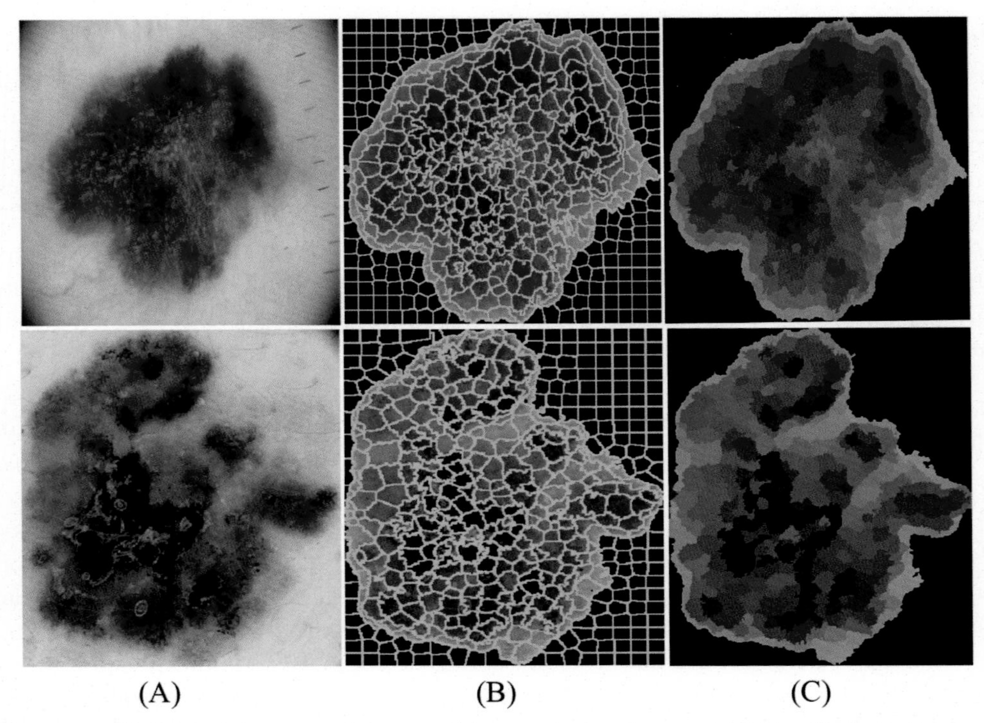

Figure 5.5 (A) Original dermoscopic images, (B) superpixel regions, and (C) identified color regions.

Output: I_c: Image with identified color regions

Step 1. Compute P, the number of pixels present in the segmented image I_s.

Step 2. Mask the input image I with the I_s to obtain the masked image I_m.

Step 3. Initialize the number of pixels N to form a superpixel.

Step 4. Determine the desired number of superpixels $S = P/N$.

Step 5. Consider I_m and S as input parameters for *SLIC* algorithm to obtain the labeled image I_L and number of superpixels S_o

Step 6. From each label L of the superpixel regions I_L, determine the maximum intensity value C_{RGB} as the intensity value having maximum number of bins in the image histogram (*hist*(I_L)). Determine the just noticeable distance D_c of the prevailing color C_{RGB} from the reference colors to obtain the image with identified color regions I_c as the following:

for $L = 1: S_o$
  $C_{RGB} = max(hist(I_L(L)))$
  $D_c = dist\ (C_{RGB}, W, R, LB, DB, BG, B)$
  If $D_c < 6$ *do*
    $I_c(L) = C_{RGB}$
  end
end

Step 7. Obtain the image with identified color regions by repeating the Step-6 for each label L.

5.3.5 Dermoscopic structure segmentation module

The DSSM subsystem determines the dermoscopic findings that correspond to the local and global structures present in the lesion area. For the development of DermESy five dermoscopic structures considered here are pigment network, dots, globules, structureless area and blue—white veil. Presence of different differential structures helps to differentiate malignant and benign lesions. Different signal processing tools have been introduced to extract those five differential structures from the dermoscopic images. Consulting with expert dermatologist the algorithms have been improvised to include expert's knowledge.

5.3.5.1 Pigment network detection

This is considered to be an important dermoscopic structure for identifying melanocytic skin lesions. This differential structure usually formed as a reticular pattern in a lighter background, is distributed in regular or irregular mesh along the lesion [18,19]. For its detection, morphological bottom-hat filtering with linear SE has been applied on the dermoscopic image masked with the corresponding segmented image. Bottom-hat filter extracts the structure having intensity darker than their surroundings. The filtered image is passed through a bank of directional filters, to obtain the connected reticular patterns in different directions.

Since, the pigment network is distributed along the lesion with different orientations, $N+1$ filters are developed. For the development of directional filter bank, the orientation θ_d has been considered as, $\theta_d \in [0, \pi]$, $d = 1, \ldots, N$. Here, 12 directions ($d = 12$) are considered with an incremental steps of $\pi/15$. The impulse response is given by [18],

$$P_{\theta_d} = G_1(x, y) - G_2(x, y). \tag{5.1}$$

where 2D Gaussian filter G_n is given as:

$$G_n(x, y) = \Bbbk \exp\left\{ -\frac{x'^2}{2\sigma_{xn}^2} - \frac{y'^2}{2\sigma_{yn}^2} \right\}, \quad n = 1, 2. \tag{5.2}$$

Here, $\Bbbk$ is the normalization constant and the values of (x', y') are related to (x, y) as,

$$x' = x\cos\theta_d + y\sin\theta_d. \tag{5.3}$$

$$y' = y\cos\theta_d - x\sin\theta_d. \tag{5.4}$$

The image is filtered by each directional filter and maximum operation is performed at each pixel (x, y) of the output image to obtain the pigment network as a grid of thin lines. In Fig. 5.6, the segmented pigment networks and corresponding regions in the lesion area are shown for two sample dermoscopic images.

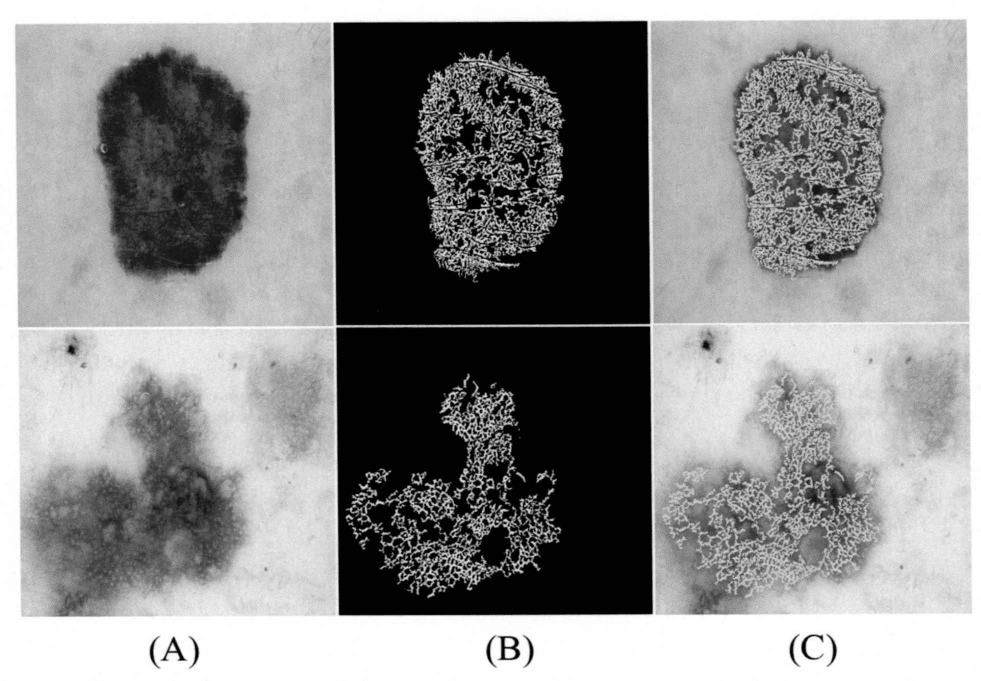

Figure 5.6 (A) Original dermoscopic images, (B) segmented pigment network, and (C) detected pigment network overlaid on the original image.

5.3.5.2 Dots and globules detection

For the categorization of melanoma, identification of dots and globules has been considered as an important step by the expert dermatologists and clinicians. Dots having small structures and globules with some circular or oval shaped area of black, brown or gray color, may be observed in center area or may be spread throughout the entire lesion. A schematic representation of the dots and globules detection technique is shown in Fig. 5.7. To exclude the dots/globules present outside the skin lesion area, the original gray scale image has been masked with the segmented binary image. The dots/globule detection algorithm is given in Algorithm 5.2. The algorithm for identification of circular structures consists of two main steps—the identification of the circular structures and the dot segmentation. In this proposed algorithm three variables, namely the threshold sensitivity (value between 0 and 1), and the minimum and the maximum radii of the circular structures have been considered.

Algorithm 5.2:
Dots and globules detection

Input: Original grayscale image I_g; *segmented image* I_s; threshold sensitivity (T_S); minimum radius R_{min} of the circular structuring element *SE* in pixels; maximum radius R_{max} of the circular structuring element *SE* in pixels.

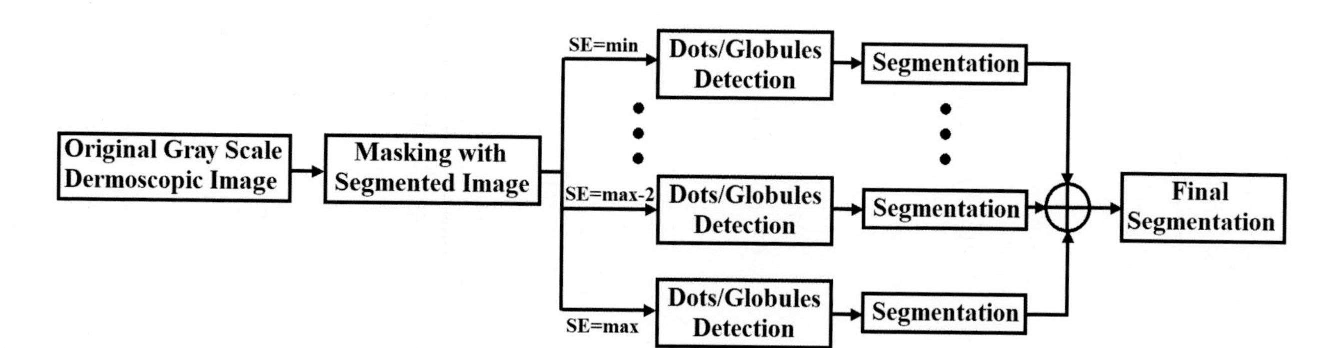

Figure 5.7 Block diagram representation of dots/globules segmentation technique.

Output: I_o: Dots and globules detected image.

Step 1. Mask the input image I_g with I_s to obtain the masked image I_m.

Step 2. Initialize the input variables as $T_S = 0.02$; $R_{min} = 2$; $R_{max} = 8$.

Step 3. Obtain the resultant image I_o with segmented dots and globules by employing varying size of the *SE* from $R_{min}$to R_{max} as:

3.1 ***for*** each radius *r* from R_{min} to R_{max} with an increment of 2 pixels ***do***:

3.2 form the circular *SE* for each radius *r*.

3.3 obtain the closed image I_c by applying morphological closing operation on I_m using circular *SE*.

3.4 apply morphological reconstruction of I_c under I_m by considering 4-connected neighboring pixels to obtain the reconstructed image I_R.

3.5 apply the morphological bottom-hat filter by subtracting I_g from I_R to obtain the image I_d with the extracted objects having smaller in size compared to *r* and intensity darker than their surroundings.

3.6 estimate the minimum intensity I_{dmin} present in the extracted regions in I_d.

3.7 estimate the maximum intensity I_{dmax} present in the extracted regions in I_d.

3.8 calculate the threshold value $T_h = T_s \times (I_{dmax} - I_{dmin}) + I_{dmin}$.

3.9 segment the objects in I_d having intensity greater than T_h to obtain the dots and globules detected image I_o.

3.10 repeat steps 3.1 to 3.9.

Step 4. Aggregate all the resultant images I_o to extract dots and globules from the input image.

For the identification of the circular structures, morphological operations have been performed with a circular kernel as a SE of varying size. The size of the SE has been varied from the earlier specified minimum radius of the circular structure to the specified maximum radius, with an increment of two-pixel distance. Morphological closing operation has been performed with each of the SE, followed by a morphological reconstruction using 4 connected neighbors. Morphological gradient operation has been performed by subtracting the original gray scale image from the reconstructed image, to identify the small circular structures with the prespecified minimum and maximum radii. In this algorithm, the minimum and the maximum radius of the dots has been considered as four pixels and 16 pixels, respectively. So, as already explained, the size of the circular kernel (SE) has been varied from a four-pixel distance to a 16-pixel distance with an increment of two pixels. The threshold sensitivity for the segmentation of dots/globules is 0.2, that is, 20% of the difference between the minimum and maximum intensity values within that particular region.

After identification of the circular structures, they have been segmented out from the original dermoscopic images using the threshold operation. First, the minimum and maximum intensity values present in those circular regions have been determined. According to the predefined threshold sensitivity, a percentage of the intensity range present in those regions has been selected as the threshold value. Finally, the segmented dots and globules have been replaced in the original gray scale image. Examples of the identified and segmented dots/globules for melanoma and nonmelanoma cases have been shown in Fig. 5.8.

5.3.5.3 Detection of structureless area of skin lesion

Few areas in dermoscopic images are devoid of any pigment network or any other structures like globules, flat or elevated areas. These are called structureless area. Their size is regarded to be at least 10% of the total lesion area. Structureless area can be hypo, hyper, or with regular pigmentation. In dermoscopic images, the structureless area has closely similar intensity variation as normal skin region. To extract the structureless area from the lesion, mathematical morphology aided technique is employed. The segmented image of the corresponding original dermoscopic image has been considered as mask to identify the lesion area from the original image. The morphological closing operation has been performed using circular SE of 4-pixel diameter on the grayscale lesion image masked with the original image. The closing operation eliminates the small objects compared to the size of the SE. Subtraction of the complement of the original image from the closed image provides a significant intensity variation between the structureless uniform region and the pigmented region of the skin lesion. From the resultant image, the threshold value has been determined by selecting the maximum intraclass variance of the lesion area with subsequent extraction of the structureless area. Fig. 5.9 depicts the original dermoscopic image and corresponding structureless area. Fig. 5.9C exhibits the distinguished intensity variations present in the structureless area and pigmented region after applying the morphological gradient operation. The extracted structureless areas have been portrayed in Fig. 5.9D.

5.3.5.4 Detection of blue–white veil of skin lesion

Blue–white veil is a focal bluish structure with an overlaying ground glass appearance. It is mainly a raised or clinically palpable component. Histologically, it represents highly pigmented melanocytes or melanophages or melanin within the dermis. A blue–white veil is usually found in melanoma but may also be present in Spitz nevus.

To extract the blue–white veil from the lesion, the input color dermoscopic image has been masked with the corresponding segmented image to confine the findings within the lesion area only. For the identification of blue–white veil, the SLIC algorithm has been considered. SLIC algorithm subdivides the ROI into smaller superpixels depending on the present color information. From each of the Superpixel region, mean intensity values of each R, G and B plane have been estimated. Measuring the distance between the mean intensity of each superpixel region and corresponding reference values as annotated by the dermatologist, the blue–white veil regions have been identified. Fig. 5.10 depicts the original dermoscopic images and its

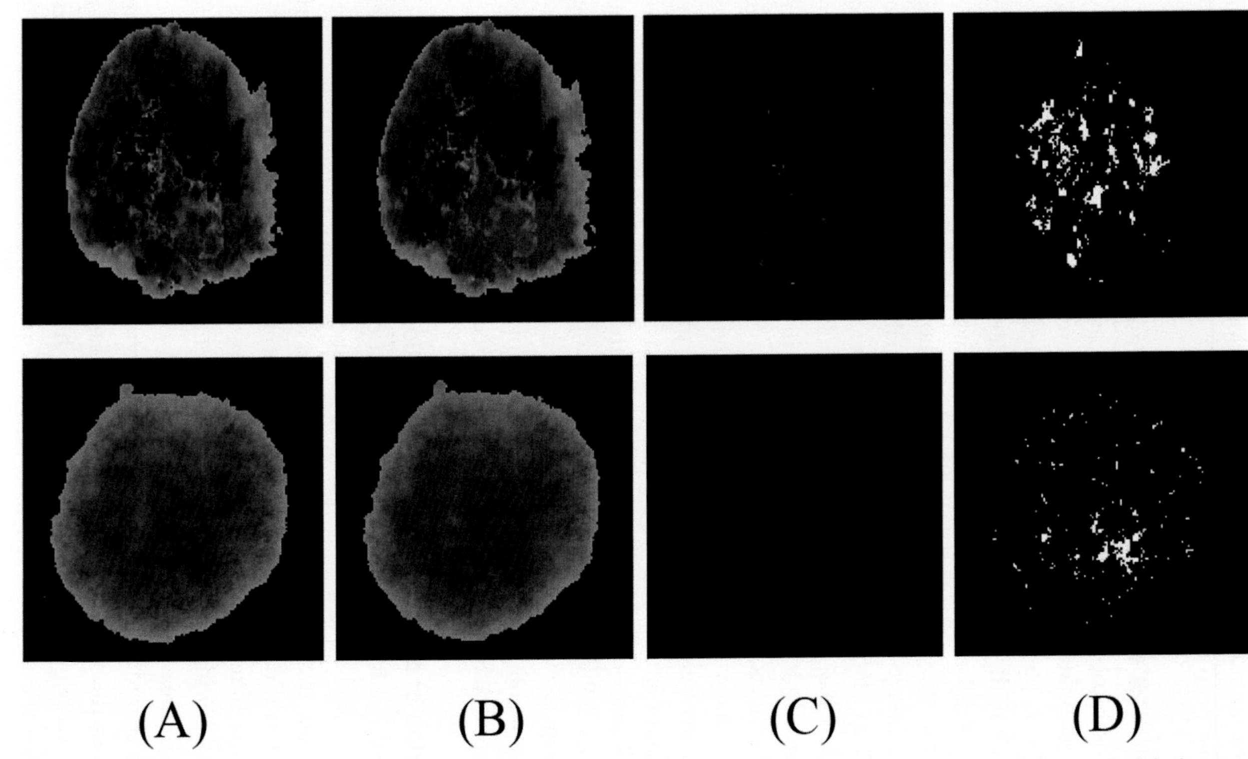

Figure 5.8 (A) Original dermoscopic images masked with the corresponding segmented image, (B) extracted dots and globule regions, and (C) detected dots and globules.

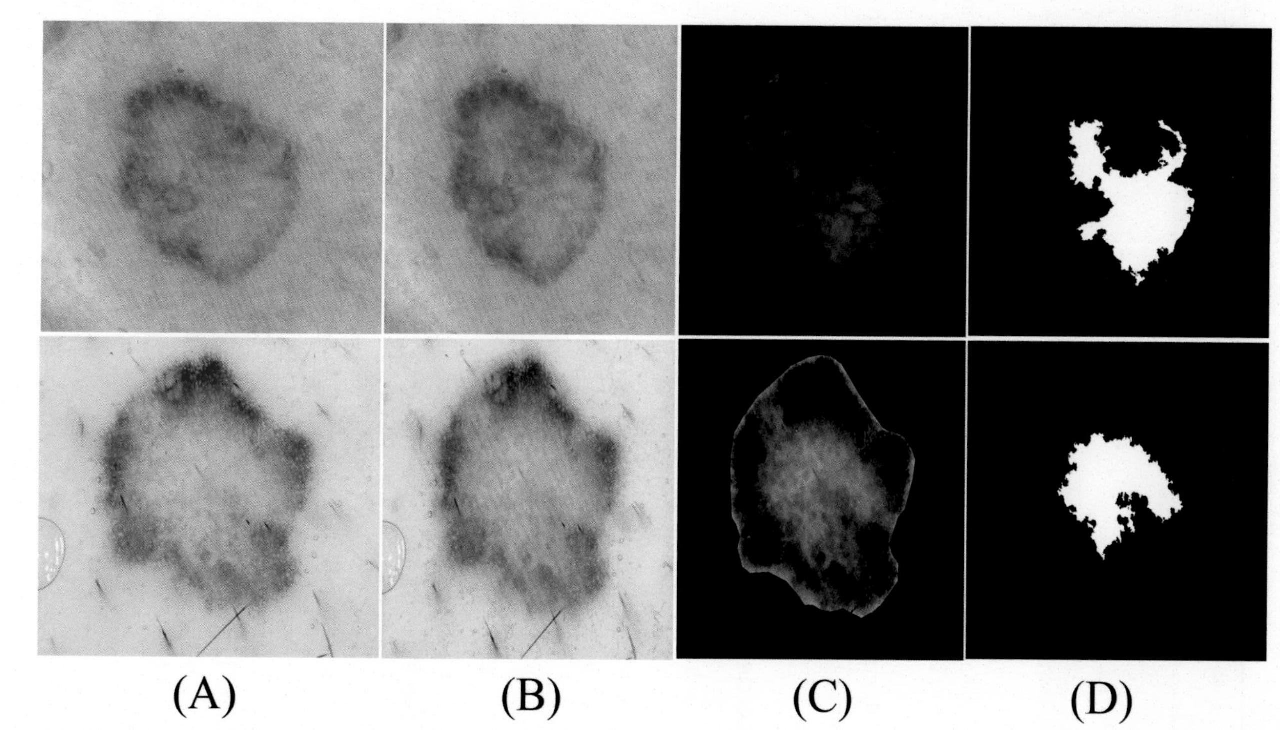

Figure 5.9 (A) Original dermoscopic images; (B) grayscale image; (C) original masked with the corresponding segmented image after morphological gradient operation; (D) extracted structureless area.

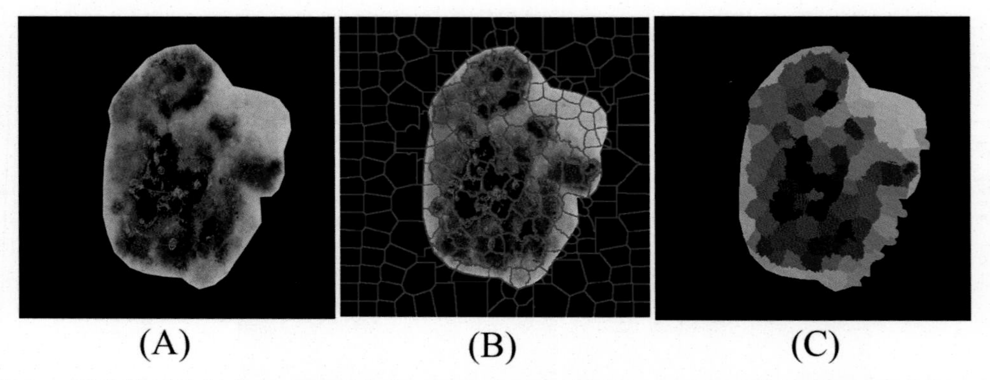

Figure 5.10 (A) Original dermoscopic images masked with the corresponding segmented image; (B) superpixel regions; (C) identified blue—white veil regions.

corresponding superpixel regions. In Fig. 5.10D blue—white veil regions have been observed according to their mean values.

5.4 Rule base design

The rule-base module has been incorporated for obtaining the TDS and finally, for the detection of the disease. It has been designed for the parameter estimation of asymmetry (A), border irregularity (B), color (C) and differential structures (D). For an input dermoscopic image, the estimated values of the ABCD attributes will act as correlated information for the explanatory subsystem module. Similarly, the estimated TDS has been used for the design of rule-based disease classification technique as an integrated part of a decision-making module.

5.4.1 Rule base design to determine "A" score

The asymmetry estimation considers shape, brightness and the color asymmetry with respect to the horizontal and vertical axes, crossing the center of gravity of the segmented lesion. To estimate the asymmetry of the lesion, an asymmetricity score has been assigned from the minimum value of 0 to the maximum value of 2. If asymmetry (considering shape, brightness, and color) is absent with respect to both the axes then the asymmetry score is zero. For single axis asymmetry, the asymmetricity score has been scaled as 0.25 for the asymmetry with respect to any one parameter (shape, brightness and color), 0.5 for any two parameters and 1 for all the parameters. Similarly, for both axes asymmetry, the asymmetricity score has been scaled as 1.25, 1.5, and 2 for the asymmetry of single parameter, any two parameters and all parameters, respectively. To estimate the shape asymmetry, the difference between the areas of two halves of the segmented lesion obtained from LSM subsystem have been evaluated along both horizontal and vertical axes. The segmented image has been rotated according to the orientation of its principal axis as illustrated in Fig. 5.11B. Two halves of the lesion and the folded bottom half have also been shown in Fig. 5.11C, D, and E, respectively. The difference between the top and folded bottom halves have been estimated by applying XOR operation on the binary segmented plane as depicted in Fig. 5.11F. Inspired by the standard ABCD rule elaborated in Ref. [6] and in consultation with dermatologist involved in this study, deviation of 2% of the estimated number of pixels along any one of the axis has been considered as the shape asymmetry. So, the shape asymmetry condition is as follows:

$$A_V = \frac{abs|A_L - A_R|}{A} \quad \text{or} \quad A_H = \frac{abs|A_T - A_B|}{A} > 0.02. \tag{5.5}$$

Here, A_V corresponds to vertical shape asymmetry determining the deviation of number of pixels between left (A_L) and right half (A_R) with respect to the total area of the lesion (A). Similarly, A_H signifies the horizontal shape asymmetry considering the top (A_T) and bottom half (A_B) of the lesion.

The difference between the average intensity of the two halves [left (L_L) and right (L_R) or top (L_T) and bottom (L_B)] along both horizontal (L_H) and vertical (L_V) axes has been estimated to evaluate the brightness asymmetry of the lesion. In reference to the previous work and the quality of the images considered in this study, the threshold value has been chosen as 3% deviation with respect to the average intensity of the lesion area (L). Therefore, the condition for brightness asymmetry has been given as:

$$L_V = \frac{abs|L_L - L_R|}{L} \quad \text{or} \quad L_H = \frac{abs|L_T - L_B|}{L} > 0.03. \tag{5.6}$$

Algorithm 5.3 Rule base to estimate shape asymmetry	**Algorithm 5.4 Rule base to estimate brightness asymmetry**	**Algorithm 5.5 Rule base to estimate color asymmetry**

If A_V && $A_H < 0.02$
$\boldsymbol{A_S = 0}$
Else
If A_V *or* $A_H > 0.02$
$\boldsymbol{A_S = 1}$
Else
If A_V && $A_H > 0.02$
$\boldsymbol{A_S = 2}$
End If
End If
End If
	If L_V &&$L_H < 0.03$
	$\boldsymbol{A_B = 0}$
	Else
	If L_V *or* $L_H > 0.03$
	$\boldsymbol{A_B = 1}$
	Else
	If L_V &&$L_H > 0.03$
	$\boldsymbol{A_B = 2}$
	End If
	End If
	End If
		If $SC_V > AC_V$&&$SC_H > AC_H$
		$\boldsymbol{A_C = 0}$
		Else
		If $SC_V < AC_V$ *or* $SC_H < AC_H$
		$\boldsymbol{A_C = 1}$
		Else
		If $SC_V < AC_V$&&$SC_H < AC_H$
		$\boldsymbol{A_C = 2}$
		End If
		End If
		End If

#H, V are horizontal and vertical axes, respectively. ##A_s, A_B and A_c are shape, brightness and color asymmetry, respectively. ###SC and AC are symmetric and asymmetric colors, respectively.

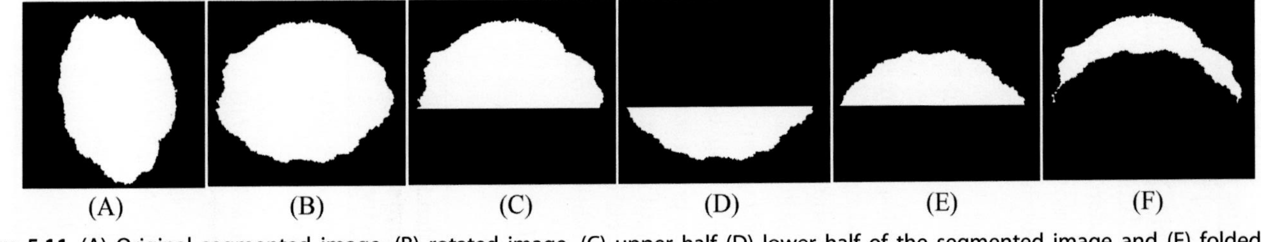

Figure 5.11 (A) Original segmented image, (B) rotated image, (C) upper half (D) lower half of the segmented image and (E) folded lower half, and (F) difference image.

The rule bases to estimate shape and brightness asymmetry are given in Algorithm 5.3 and Algorithm 5.4, respectively.

To obtain color asymmetry present in the two halves of the lesion area along horizontal axis and also along vertical axis, each half of the lesion has been subdivided into smaller superpixel regions according to the color information. For the superpixel generation, the SLIC algorithm has been implemented. The number of superpixels has been determined as the ratio of the total number of ROI pixels in the corresponding half and the minimum number of pixels to form a superpixel. Here, 32 numbers of pixels have been selected to form a superpixel. The number of superpixels having same median value has been considered as the symmetric color regions. The symmetric and asymmetric color regions in both half of the lesion area along each of the axes have been determined. The lesion has been considered to be color asymmetric along any axis if the number of asymmetric color regions is more than the symmetric color regions along that axis. The entire rule base for the estimation of color asymmetry has been shown in Algorithm 5.5. The final rule base to determine the asymmetry of the lesion has been shown in Algorithm 5.6.

5.4.2 Rule base design to estimate "B" index

The border detected image obtained from BDM subsystem has been subdivided into eight equal segments. In ABCD rule, the border irregularity score between 0 and 8 has been estimated by evaluating the irregularity present in each of the eight segments. If none of the segment has irregularity, then the minimum score of 0 has been considered and a score of 1 has been assigned for the irregularity of each segment up to the maximum score of 8. Here, Katz fractal dimension [20] has been determined from the border series of each segment. The fractal dimension of a planar curve is:

$$D = \frac{\log(P)}{\log(d)}. \tag{5.7}$$

where P is the total length of the curve and d is the diameter or planar extent of the curve [20]. The total length of the curve is determined as the sum of the distances between successive points of the curve. The diameter d is the maximum distance between the first sample and all the subsequent samples in the time series.

Considering the average step a, that is, the average distance between the successive points of the time series, number of steps (n) in the time series has been determined to be $n = P/a$.

For n number of steps of the curve of length (P) and diameter (d), the Katz fractal dimension [20] has been determined as

$$K_D = \frac{\log(n)}{\log(d/P) + \log(n)}. \tag{5.8}$$

The skin lesions are commonly circular in nature. Considering the fractal dimension of a circle as the reference, the irregularity of the lesion has been determined from the deviation of the fractal dimension of that structure with respect to the reference value. The deviation of more than 10% from the reference has been considered to label the segment as asymmetric.

5.4.3 Rule base design to determine "C" score

The CIEM subsystem has provided the information regarding the presence of six different dermatologically significant colors in the lesion area. To calculate the *TDS*, the score has been assigned as zero for the absence of any of the six colors and maximum score of six for the presence of all of the six colors. For the presence of each distinguished color, a score of one has been assigned.

5.4.4 Rule base design to obtain "D" score

Here, DSSM subsystem has been developed to extract five dermoscopic structures as pigment network, dots, globules, structureless area and blue–white veil. Minimum score of zero to maximum of five has been assigned for the absence of any structure to the presence of all the five structures, respectively. According to the standard

Algorithm 5.6 Rule base for estimating asymmetry of the lesion

If A_S && A_B && $A_C = 0$
    $\mathbf{A = 0}$
  Else
    If A_S *or* A_B *or* $A_C = 1$
      $\mathbf{A = 0.25}$
      If A_S&A_B *or* A_S&A_C *or* A_C&$A_B = 1$
        $\mathbf{A = 0.5}$
      Else
        If A_S&& A_B&& $A_C = 1$
          $\mathbf{A = 1}$
      End If
    End If
  Else
  If A_S *or* A_B *or* $A_C = 2$
    $\mathbf{A = 1.25}$
    If A_S&A_B *or* A_S&A_C *or* A_C&$A_B = 2$
      $\mathbf{A = 1.5}$
    Else
      If A_S&& A_B&& $A_C = 2$
        $\mathbf{A = 2}$
      End If
    End If
  End If
  End If
End If

ABCD rule, for the presence of each structure, a score of one has been assigned. For pigment network, dots, globules, and structureless area, size more than 10% of the entire lesion area has been considered for existence of the structure. The irregular or localized distribution of pigment network, dots, and globules in the lesion area is more indicative towards melanoma than globally distributed structures. Considering these dermoscopic findings, here an improvisation on the ABCD rule is proposed, by adding a score of 0.5 for each of the locally distributed structure (pigment network, dots and globules) satisfying the following conditions:

- *Absent structure*: if the dermoscopic structure (DS) has an area of less than 10% of the total area of the lesion (A_L), then the structure has been considered to be absent [15].

$$Area(DS) < 0.1 \times A_L. \tag{5.9}$$

- *Local structure*: if the dermoscopic structure contains more than 10% but less than 50% of the total pixels of the lesion area, then the structure has been considered to be locally distributed along the lesion area.

$$0.1 \times A_L < Area(DS) < 0.5 \times A_L. \tag{5.10}$$

- *Global structure*: if the dermoscopic structure contains more than 50% of the total pixels of the lesion area, then the structure has been considered to be global structure.

$$Area(DS) > 0.5 \times A_L. \tag{5.11}$$

The presence and absence of the structureless area has been considered by assigning a score of 1 and 0, respectively. The presence of the structureless area is predominant if it occupies more than 10% of the entire lesion area and is considered to be absent otherwise. The spatial information of the structureless area does not have any significant effect on lesion identification. Therefore, no further improvisation is introduced for the estimation of the overall "D" score considering the spatial properties of the structureless area. Blue–white veil can be found in both malignant and benign lesions. If the blue–white veil area covers a significant region (more than 30% of the entire lesion area), it is considered as feature for benign lesion. Under such condition, a score of 1 has been added to the entire "D" score for blue–white veil with more than 10% area of the entire lesion. Similarly, for the structure occupying more than 30% of the entire lesion area, value of 0.5 is subtracted from the entire "D" score.

The entire rule base to determine the *D* score is shown in Algorithm 5.7.

Algorithm 5.7 Rule base for determining the *D* score

If $A_{PN}A_DA_GA_{SL}A_{BWV} < 0.1 \times A_L$

  $\boldsymbol{D = 0}$

Else

  If $A_{PN} or A_D or A_G or A_{SL} > 0.1 \times A_L$

    $\boldsymbol{D = 1}$

If$\mathcal{A}_{PN}$ *or* $\mathcal{A}_D$ *or* $\mathcal{A}_G < 0.5 \times \mathcal{A}_L$
D = D + 1
Else
If $\mathcal{A}_{PN}$ *or* $\mathcal{A}_D$ *or* $\mathcal{A}_G > 0.5 \times \mathcal{A}_L$
D = D + 0.25
End If
End If
End If
If $\mathcal{A}_{PN}\mathcal{A}_D$ *or* $\mathcal{A}_D\mathcal{A}_G$ *or* $\mathcal{A}_{PN}\mathcal{A}_G$ *or* $\mathcal{A}_{SL}\mathcal{A}_{PN}$ *or* $\mathcal{A}_{SL}\mathcal{A}_D\mathcal{A}_{SL}\mathcal{A}_G > 0.1 \times \mathcal{A}_L$
D = 2
If $\mathcal{A}_{PN}\mathcal{A}_D$ *or* $\mathcal{A}_D\mathcal{A}_G$ *or* $\mathcal{A}_{PN}\mathcal{A}_G < 0.5 \times \mathcal{A}_L$
D = D + 1
Else
If $\mathcal{A}_{PN}\mathcal{A}_D$ *or* $\mathcal{A}_D\mathcal{A}_G$ *or* $\mathcal{A}_{PN}\mathcal{A}_G > 0.5 \times \mathcal{A}_L$
D = D + 0.25
End If
End If
End If
If $\mathcal{A}_{PN}\mathcal{A}_D\mathcal{A}_G$ *or* $\mathcal{A}_{PN}\mathcal{A}_D\mathcal{A}_{SL}$ *or* $\mathcal{A}_{SL}\mathcal{A}_D\mathcal{A}_G > 0.1 \times \mathcal{A}_L$
D = 3
If$\mathcal{A}_{PN}\mathcal{A}_D\mathcal{A}_G < 0.5 \times \mathcal{A}_L$
D = D + 1
Else
If $\mathcal{A}_{PN}\mathcal{A}_D\mathcal{A}_G > 0.5 \times \mathcal{A}_L$
D = D + 0.25
End If
End If
End If
If $\mathcal{A}_{PN}\mathcal{A}_D\mathcal{A}_G\mathcal{A}_{SL} > 0.1 \times \mathcal{A}_L$
D = 4
If$\mathcal{A}_{PN}\mathcal{A}_D\mathcal{A}_G < 0.5 \times \mathcal{A}_L$
D = D + 1
Else
If $\mathcal{A}_{PN}\mathcal{A}_D\mathcal{A}_G > 0.5 \times \mathcal{A}_L$
D = D + 0.25
End If
End If
End If
If $\mathcal{A}_{BWV} > 0.1 \times \mathcal{A}_L$
D = D + 1
If $\mathcal{A}_{BWV} > 0.3 \times \mathcal{A}_L$
D = D−0.5
End If
End If
End If

#$\mathcal{A}_{PN}, \mathcal{A}_D, \mathcal{A}_G, \mathcal{A}_{SL}$, $\mathcal{A}_{BWV}$ *and* $\mathcal{A}_L$ are area of pigment network, dots, globules, structureless area, blue−white veil and entire lesion, respectively.

5.4.5 Calculation of total dermoscopic score to design the classifier

Following the standard ABCD rule of dermoscopy [10], TDS is estimated by multiplying the ABCD attributes with the coefficients 1.3, 0.1, 0.5 and 0.5, respectively, as given here:

$$TDS = A \times 1.3 + B \times 0.1 + C \times 0.5 + D \times 0.5. \tag{5.12}$$

To classify the diseases based on TDS, the lesions have been categorized into three classes as benign, suspicious, and malignant melanoma. The low TDS determines the benign lesions (TDS < 4.75) while an intermediate score between 4.75 and 5.45 is interpreted as suspicious lesions. A high value of TDS score is considered as a malignant lesion. Here, the threshold values for the differentiation of malignant, benign, and suspicious lesions have been selected following the standard ABCD rule of dermoscopy. The rule base for the classifier has been given as Algorithm 5.8. Using the rule base, the final decision regarding the disease class and associated findings have been obtained by the inference engine.

Algorithm 5.8 Rule base for classifier

If TDS < 4.75
Benign_lesion
Else
If 4.75 < TDS < 5.45
Suspicious
Else
If TDS > 5.45
Malignant melanoma

5.5 Results and discussion

5.5.1 Dataset and experimental setup

The expert system for skin disease diagnosis has been evaluated using the official database International Skin Imaging Collaboration (ISIC) challenge 2016, 2017, 2018 dataset [21] and PH2 dataset [22]. The database has 6669 dermoscopic images, comprising of histopathologically confirmed lesions, malignant melanoma (2209), nevus (3430), basal cell carcinoma (BCC) (586), SK (419) and squamous cell carcimona (SCC) (226). The entire dataset is categorized as malignant, benign, and suspicious lesions. The dermoscopic images annotated as malignant, including malignant melanoma, malignant BCC (67), SK (1) and SCC (22) are considered for the identification of malignant lesions. The benign category is comprised of nevus (3422), SK (418) and benign BCC (1) lesions. Here, some of the identified lesions are not categorized as benign and malignant. Such dermoscopic images considered as suspicious lesions are BCC (518), SCC (204) and nevus (8). Among 6669 number of dermoscopic images, 2116 images have been annotated with differential structures (pigment network, dots, and globules). Optimum parameter/ threshold values for the estimation of dermoscopic scores, as mentioned in previous section, have been achieved at after thorough experimentations.

5.5.2 Performance evaluation of lesion segmentation module subsystem

The performance of the LSM subsystem has been determined by the quantitative estimation of the similarity between the segmented lesion and its corresponding GT image. The GT images have been annotated by an expert dermatologist. The similarity measure between the segmented images and corresponding GTs of the sample dermoscopic images have been shown in Fig. 5.12. In Fig. 5.12, images in the second column correspond to the GT images while the third column represents the segmented images obtained using LSM subsystem. In Fig. 5.12D the pink color has indicated the amount of dissimilarity between the segmented image and the corresponding GT. From the resultant image, it has been observed that the reported morphological segmentation algorithm has segregated the ROI with much closer approximation to the GT images. The lesions with complex structure and various shapes have been segmented using the morphology aided segmentation technique with circular SE. The selection of circular SE has also improved the segmentation performance and makes the algorithm robust. The goodness of segmentation has been evaluated by estimating the pixel-level sensitivity (*Sen*), specificity (*Spc*) and accuracy (*Acc*) along with the similarity measure indices as Jaccard similarity index (JSI) and dice similarity coefficient (DSC).

The segmentation performance indices of the entire dataset considered in this study has been tabulated in Table 5.1. The estimated performance indices from Table 5.1 describe the goodness of the segmentation algorithm implemented in LSM subsystem for the segmentation of wide varieties of lesions. The table shows the minimum and maximum values of the performance indices obtained from the entire dataset. Minimum values of JSI and DSC correspond to the significant deviation of the segmented image and corresponding GT image. Similarly, the maximum values indicate closer similarity between the resultant image and its corresponding GT. However, the sufficiently large average values of performance indices have testified the closeness of the segmented image and its corresponding GT for maximum number of images in the dataset. Therefore, the acceptable segmentation performance indices ensure the accurate segmentation of small as well as complex structures of various skin lesions.

5.5.3 Comparative performance analysis of the lesion segmentation module subsystem and state-of-the-art techniques

The performance of the LSM of the expert system has been compared with the state-of-the-art techniques applied on different datasets having different number of images, and the results are given in Tables 5.2 and 5.3, respectively. In this study, all the

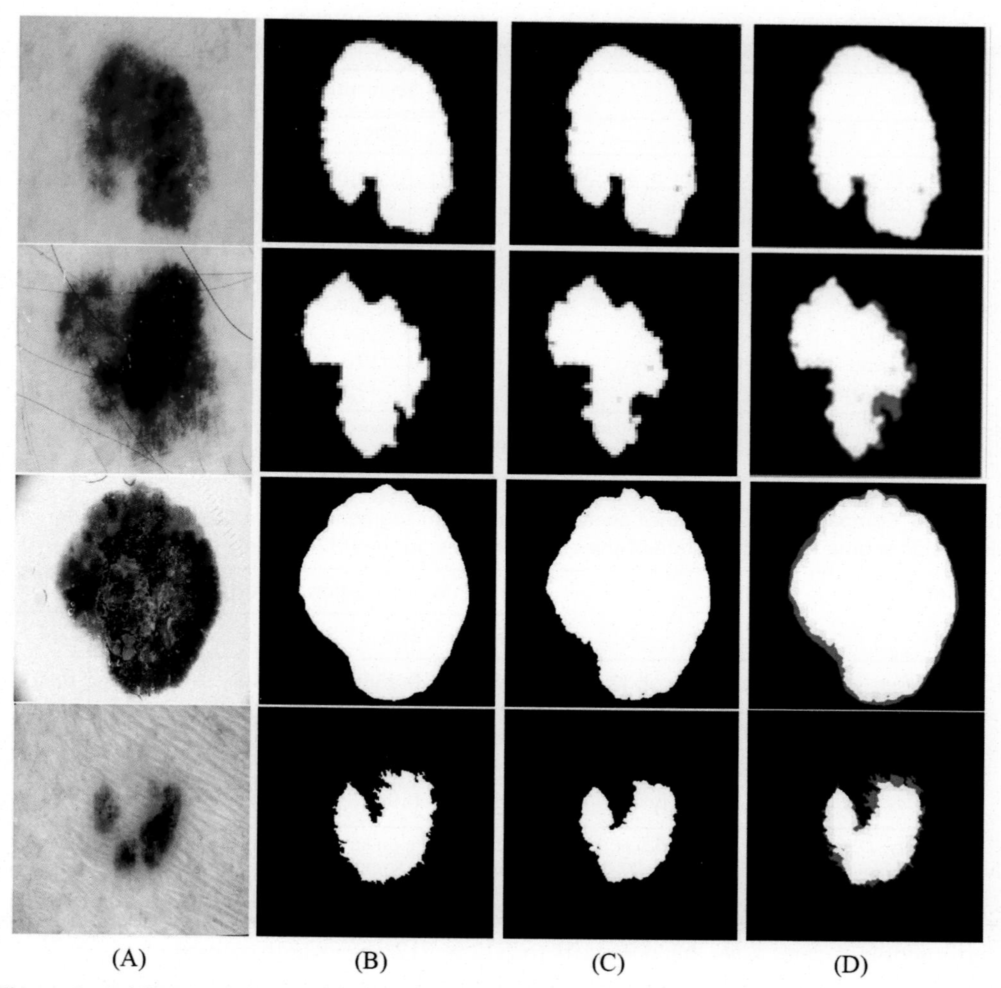

Figure 5.12 (A) Original images, (B) segmented ground truth (GT) images, (C) segmented images using lesion segmentation module, and (D) similarity measure of the segmented image with respect to the GT image.

Table 5.1 Performance of the lesion segmentation module subsystem.

Data base	Values	Segmentation performance indices				
		Sen	Spc	Acc	JSI	DSC
ISIC and PH2 dataset	Minimum	0.606	0.626	0.710	0.658	0.662
	Maximum	0.997	0.996	0.989	0.897	0.927
	Average	0.917	0.974	0.962	0.834	0.884

Acc, Accuracy; *DSC*, dice similarity coefficient; *ISBI*, international symposium on biomedical imaging; *ISIC*, International Skin Imaging Collaboration; *JSI*, Jaccard similarity index; *Sen*, sensitivity; *Spc*, specificity.

Table 5.2 Comparative performance analysis of the lesion segmentation module subsystem for skin lesion segmentation and state-of-the-art techniques on the ISIC challenge dataset.

Dataset	Method	Segmentation performance indices				
		Sen	Spc	ACC	JSI	DSC
ISBI/ISIC 2017 dataset (2000 training images, 600 testing images)	Al-masni et al. [23]	0.8540	0.9669	0.9403	0.7711	0.8708
	Bi et al. [24]	0.8620	0.9671	0.9408	0.7773	0.8566
	Shan et al. [25]	0.8382	0.9865	0.9371	0.7634	0.8456
	Xie et al. [26]	0.8700	0.9640	0.9380	0.7830	0.8620
	Present study	0.9120	0.9750	0.9480	0.8120	0.8680
ISBI/ISIC2016 dataset (900 training images, 379 testing images)	Bi et al. [24]	0.9311	0.9605	0.9578	0.8592	0.9177
	Xie et al. [26]	0.870	0.964	0.938	0.918	0.858
	Present study	0.946	0.976	0.961	0.874	0.914

Acc, Accuracy; *DSC*, dice similarity coefficient; *ISBI*, international symposium on biomedical imaging; *ISIC*, International Skin Imaging Collaboration; *JSI*, Jaccard similarity index; *Sen*, sensitivity; *Spc*, specificity.

Table 5.3 Comparative performance analysis of the lesion segmentation module subsystem for skin lesion segmentation and state-of-the-art techniques on the PH2 dataset.

Dataset	Method	Segmentation performance indices				
		Sen	Spc	ACC	JSI	DSC
PH2 dataset (200 images for testing)	Al-masni et al. [23]	0.9372	0.9565	0.9508	0.8479	0.9177
	Xie et al. [26]	0.963	0.942	0.949	0.857	0.919
	Shan et al. [25]	0.9477	0.9628	0.9363	0.8351	0.9026
	Bi et al. [24]	0.9623	0.9452	0.9530	0.8590	0.9210
	Present study	0.958	0.982	0.968	0.863	0.912

Acc, Accuracy; *DSC*, dice similarity coefficient; *ISBI*, international symposium on biomedical imaging; *ISIC*, International Skin Imaging Collaboration; *JSI*, Jaccard similarity index; *Sen*, sensitivity; *Spc*, specificity.

dermoscopic images have been resized to 512×512 pixels and converted to grayscale form. For the development of morphological filters and segmentation algorithm, circular SE has been chosen with a size of eight-pixel diameter. Therefore, for the segmentation of the skin lesions, 512×512 size of grayscale images with 8-pixel diameter circular SEs have been considered as the setting for LSM. The performance of the reported segmentation algorithm has been compared with the state-of-the-art techniques on ISIC challenge 2017 dataset in Table 5.2. It can be observed from Table 5.2 that a deep full resolution convolutional neural network for the segmentation of skin lesions as reported in Ref. [23] has achieved 0.7711 JSI, 0.8708 DSC, and 0.9403 accuracy on 600 test images of ISIC challenge 2017 dataset. In Ref. [24], 0.9408 accuracy, 0.7773 JSI, and 0.8566 DSC have been obtained for the segmentation of the skin lesions employing a deep class-specific learning approach. Performing fully convolutional neural network and dual path network technique for skin lesion segmentation from dermoscopic images by Shan et al. [25], it has been reported that the segmentation performance indices are 0.9371 of pixel-level accuracy, 0.7634 JSI, and 0.8456 DSC. High-resolution convolutional neural network for the segmentation of the skin lesions has been implemented in Ref. [26] achieving 0.9380 accuracy, 0.7830 JSI, and 0.8620 DSC on ISIC 2017 dataset. To compare the present study with the above mentioned techniques, segmentation performance indices have been evaluated on the same 600 dermoscopic images considered by the other authors. In the study reported in the present chapter, the mathematical morphology aided segmentation technique has segregated the lesion area from the dermoscopic images with acceptable performance indices of 0.912 sensitivity, 0.975 specificity, 0.948 accuracy, 0.812 JSI, and 0.868 DSC. Table 5.2 reveals that the methodology illustrated here has outperformed the state-of-the-art segmentation techniques on ISIC challenge 2017 dataset. The performances of the works by Bi et al. [24] and Xie et al. [26] have also been tested on ISIC challenge 2016 dataset. The segmentation technique reported by Bi et al. [24] has achieved 0.9311 sensitivity, 0.9605 specificity, 0.9578 accuracy, 0.8592 JSI, and 0.9177 DSC, evaluated on 379 dermoscopic images. The research work by Xie et al. [26] has reported the segmentation performance indices of 0.87 sensitivity, 0.964 specificity, 0.938 accuracy, 0.918 JSI, and 0.858 DSC on the ISIC challenge 2016 dataset. To compare the study on morphology aided segmentation technique with the state-of-the-art techniques, the performance indices have been evaluated on the same set of 379 dermoscopic images of ISIC challenge dataset and it has obtained 0.946 sensitivity, 0.976 specificity, 0.961 accuracy, 0.874 JSI, and 0.914 DSC.

The comparative performance analysis of the present study with some of the published work on PH2 dataset has been tabulated in Table 5.3. The research group led by Al-masni et al. [23] have evaluated the segmentation performance of the reported work on 200 dermoscopic images of PH2 dataset and achieved 0.9372 sensitivity, 0.9565 specificity, 0.9508 accuracy, 0.8479 JSI and 0.9177 DSC. The work by Bi et al.[24] has tested the segmentation performance on 200 dermoscopic images of PH2 dataset with 0.9623 sensitivity, 0.9452 specificity, 0.9530 accuracy, 0.8590 JSI and 0.9210 DSC. On similar dataset, the segmentation technique employing fully convolutional neural network and dual path based approach by Shan et al. [25] has achieved 0.9477 sensitivity, 0.9628 specificity, 0.9363 accuracy, 0.8351 JSI and 0.9026 DSC. In Ref. [26], 0.963 sensitivity, 0.942 specificity, 0.949 accuracy, 0.857 JSI and 0.919 DSC have been obtained by using high-resolution convolutional neural network-based segmentation technique applied on dermoscopic images of PH2 dataset. To compare the reported segmentation algorithm implemented in LSM subsystem of the DermESy with the state-of-the-art techniques tabulated in Table 5.3, similar set of dermoscopic images have been considered. Tested on the 200 dermoscopic images of PH2 dataset, the reported segmentation algorithm has achieved 0.958 sensitivity, 0.982 specificity, 0.968 accuracy, 0.863 JSI and 0.912 DSC. Table 5.3 demonstrates that the present study significantly improves the segmentation performance of the other reported segmentation algorithms, tested on PH2 dataset.

5.5.4 Performance evaluation of dermoscopic structures segmentation module subsystem

5.5.4.1 Performance evaluation of pigment network detection system

Performance of the reported pigment network detection system has been evaluated by considering the relevant GT images provided in the ISIC archive database. The corresponding pigment network regions of a skin lesion area are marked in the GT images by the dermatologists. Therefore, the similarity measure between the detected pigment network and corresponding GT describes the effectiveness of pigment network detection algorithm. Performance of the algorithm has been evaluated by estimating pixel-level sensitivity (Sen), specificity (Spc), accuracy (Acc), similarity coefficient of JSI, and DSC as given in Table 5.4. From the table, the minimum values of performance indices correspond to the variations of the detected region from the GT and the maximum values determines the maximum similarity between the detected region and corresponding GT. Considering the lesions containing pigment network, the average values of the performance indices have been determined. The acceptable values of the performance indices demonstrate the identification of pigment network for maximum number of images with higher degree of similarity with the GTs.

5.5.4.2 Performance evaluation of dots and globules detection system

To evaluate the performance of the morphological filter aided dots and globules detection algorithm, corresponding GT images given in ISIC archive dataset have been considered. As tabulated in Table 5.5, pixel-level sensitivity (Sen), specificity

Table 5.4 Performance evaluation of pigment network detection system of the DSSM (dermoscopic structure segmentation module) subsystem.

Structures	Values	Segmentation performance indices				
		Sen	Spc	Acc	JSI	DSC
Pigment network	Minimum	0.562	0.545	0.556	0.591	0.572
	Maximum	0.953	0.961	0.963	0.876	0.869
	Average	0.914	0.925	0.913	0.806	0.811

Acc, Accuracy; *DSC*, dice similarity coefficient; *JSI*, Jaccard similarity index; *Sen*, sensitivity; *Spc*, specificity.

Table 5.5 Performance evaluation of dots and globules detection system of the DSSM subsystem.

Structures	Values	Segmentation performance indices				
		Sen	Spc	Acc	JSI	DSC
Dots and globules	Minimum	0.581	0.569	0.603	0.537	0.541
	Maximum	0.981	0.977	0.983	0.881	0.863
	Average	0.913	0.935	0.930	0.865	0.871

Acc, Accuracy; *DSC*, dice similarity coefficient; *JSI*, Jaccard similarity index; *Sen*, sensitivity; *Spc*, specificity.

(Spc), and accuracy (Acc) are determined and the degree of similarity between the segmented dots, globules regions and corresponding GT are determined by JSI and DSC. Table 5.5 shows the minimum, maximum, and average values of performance indices for dots and globules detection. The acceptable values of JSI and DSC describe the detection of dots and globules with high degree of similarity.

5.5.5 Assessment of the proposed system in automatic diagnosis

Reported DermESy has differentiated the malignant and benign lesions in accordance to the evaluated TDSs. As per the dermoscopic properties and corresponding ABCD attributes, the suspicious lesions have been identified which can be used for further analysis of biopsy for confirmation. Depending on the identification of actual disease classes the confusion matrix has been represented in Fig. 5.13A. According to the implemented ABCD rule, the lesions having TDS in between 4.75 and 5.45 have been considered to be suspicious lesions. The confusion matrix as shown in Fig. 5.13A has represented the number of lesions correctly identified as melanoma and benign, as well as the suspicious lesions. The confusion matrix reveals that the reported expert system has correctly identified 2246 malignant melanoma cases with 15 images as suspicious. Similarly, 3763 numbers of benign lesions have been identified correctly with 42 images in suspicious category. Among 730 images of suspicious category have been categorized as malignant (243), benign (312), and suspicious (175) lesions. The characterization of these suspicious lesions in benign and malignant category may further help the dermatologists in their final decision-making. All the dermoscopic images considered in this expert system have been given to an expert dermatologist for the characterization of the lesions in benign, malignant, and suspicious categories, considering their ABCD criterion. The dermatologist has been asked to evaluate the TDS for each of the dermoscopic images and to provide a concluding remark on dermoscopic findings. Performance of the dermatologist in identification of malignant and benign lesions has been represented by the confusion matrix shown in Fig. 5.13B. From the confusion matrix, it has been observed that dermatologist has correctly

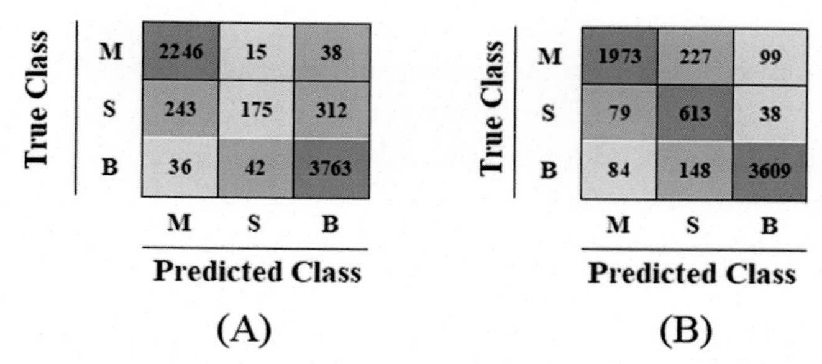

Figure 5.13 (A) Confusion matrix of DermESy, (B) confusion matrix of expert's diagnosis. (*M*, malignant; *S*, suspicious; *B*, benign).

identified 1973 melanoma images and 3609 benign lesions. Therefore, the results reveal that a significant number of benign lesions have been segregated by the dermatologist compared to the malignant lesion due its complex visual appearance. The confusion matrix corresponding to the dermatologist's diagnosis has also expressed that a small number of suspicious lesions have been further identified as malignant and benign lesion by the expert. Results of the confusion matrices justify the improvement of diagnosis by DermESy with appropriate evaluation of ABCD parameters of the dermoscopic rule. The quantitative analysis helps to identify the malignant and benign lesions from the suspicious lesion category. Therefore, diagnosis of DermESy can be considered as a second opinion by the dermatologist for improved diagnosis and decision-making in further biopsy of the lesions. The skin disease identification performance of the expert system has been summarized based on three classification baseline parameter, namely sensitivity (SE), specificity (SP) and correct classification accuracy (ACC) as the following.

$$\text{SE} = \text{No. of true positive (TP) skin disease samples} \\ /\text{No. of all skin disease samples classified as positive (TP + FN)}$$

$$\text{SP} = \text{No. of true negative (TN) skin disease samples} \\ /\text{No. of all skin disease samples classified as negative (TN + FP)}$$

$$\text{ACC} = \text{No. of correctly classified skin disease samples (TP + TN)} \\ /\text{Total no. of skin disease samples (TP + TN + FP + FN)}$$

where TP, positive samples classified as positive; TN, negative samples classified as negative; FP, negative samples misclassified as positive; FN, positive samples misclassified as negative.

The comparative analysis of the dermatologist's performance with the expert system has been depicted in Fig. 5.14. The expert system has differentiated the malignant and benign lesions with 97.69% sensitivity, 97.97% specificity and 97.86% accuracy. Dermatologist has detected the malignant and benign lesions with 85.82% of sensitivity, 93.96% specificity and 90.91% accuracy. From Fig. 5.14, it has been observed that the quantification of the dermoscopic findings has improved the skin disease identification performance significantly. The proper estimation of the TDS has reduced the number of suspicious lesions leading to the enhanced identification accuracy for deciding the further course of treatment.

The methodology illustrated in Ref. [27], have tested the diagnosis of malignant melanoma from 165 pigmented skin lesions with a Kappa value, a score measuring the agreement between two or more human evaluators of 0.96 employing ABCD rule. The results have been verified with three experts having different spans of experience. However, the performance of the system has been evaluated on a very small dataset,

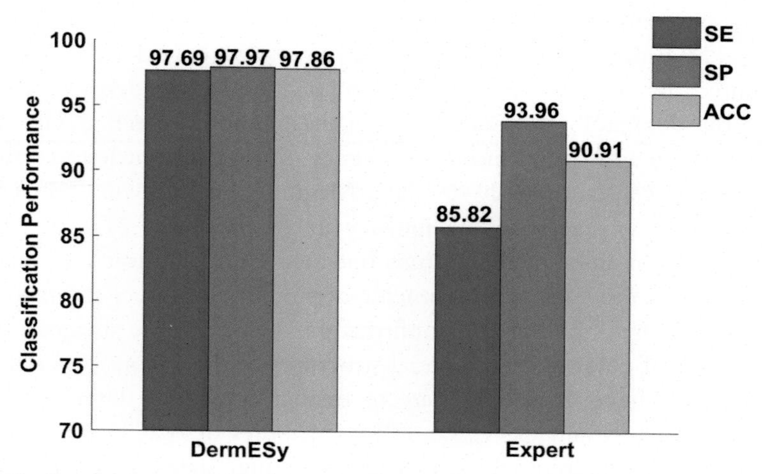

Figure 5.14 Performance comparison between DermESy and expert's diagnosis.

with minimum variability in samples in comparison with this present study. An ABCD rule-based system has been developed by Smaoui and Bessassi [28] estimating morphological features, colors, and diameter of the lesions. The work has reported 92% classification accuracy for 40 dermoscopic images. The present study has improvised the attributes of the ABCD rule and considering the differential structures 97.87% accuracy has been achieved for 6669 dermoscopic images. As reported in Ref. [6], 94% accuracy has been achieved for the differentiation of malignant and benign skin lesions on the basis of ABCD rule. In that proposed methodology, the dermoscopic structures have not been segmented for proper visual inspection by the dermatologists. For the estimation of ABCD attributes, the spatial information of the dermoscopic structures have not been considered. However, in this present study, the TDS has been estimated with improvised ABCD attributes considering the spatial information of the dermoscopic structures. The performance of this study has also been verified with the diagnostic performance indices with an expert dermatologist. The skin disease diagnosis technique described in Ref. [15] has developed a knowledge based convolutional neural network model for skin disease diagnosis. This technique has not incorporated a proper segmentation method for extraction of features related to the structural properties of the lesions. An appropriate visualization module for the identification of dermoscopic findings has not been introduced in the model. However, DermESy is developed considering a segmentation module for accurate segmentation of skin lesion area to estimate asymmetry index of the lesion and also to monitor further spreading of the disease. In this model, an explanatory subsystem is introduced to recognize the dermoscopic finding related to the ABCD rule of dermoscopy and to correlate the decision with an expert dermatologist. Computer aided classification system has been developed to differentiate suspicious and nonsuspicious pigmented lesions using ABCD features by Birkenfeld et al. [29]. The authors have considered 1759 wide-field images, acquired by a consumer grade camera instead of dermoscopic images. According to the dermatologist associated with the present research group, naked eye examination of the lesion shows more invasiveness of a malignant lesion compared to a benign one as the cells are rapidly dividing. Magnified images were taken into consideration to raise the suspicion of malignancy in a better way. "D" of ABCD rule stands for diameter or differential structure obtained from dermoscopic images only. For clinical images, consideration of diameter is enough to suspect the malignancy of the lesion followed by a biopsy for confirmation. Dermoscopic images provide in-depth visualization of color and differential structures of the lesions. Considering this, dermoscopic images have been used here to extract color and differential structures from the lesion area to evaluate "C" and "D" attributes of ABCD rule. The authors [29] have extracted different texture features using state-of-the-art techniques. However, literature suggests that the information regarding the presence of differential structures in skin lesions cannot be assessed through existing texture analysis tools. In the study presented here, various image processing tools have been employed to extract the differential structures present in the lesion and accordingly the ABCD rule has been improvised by considering not only the existence of the structures but also their spatial properties. The reported DermESy has differentiated malignant and benign lesions with 97.86% accuracy compared to the accuracy of 75.9% reported in Ref. [29]. The work by Isasi et al. [30] focuses primarily on the development of different pattern recognition algorithms to extract globular, reticular and blue-veil patterns from the skin lesions. Apart from extracting such differential structures, an indigenous expert system has also been designed in the present study for proper estimation of the ABCD scores with a rule-based classifier to provide a second opinion for improved diagnosis and monitoring of the disease. The reported algorithms [30] have been employed in 160 dermoscopic images whereas the present expert system considers 6669 dermoscopic images for the extensive analysis of all ABCD attributes and the diagnosis of the diseases. An ABCD rule-based pigmented skin lesion identification technique has been developed by Mabrouk et al. [31]. The authors have used ABCD feature based SVM classification technique for the differentiation of malignant and benign lesions from 320 dermoscopic images. In this reported DermESy system, TDS has been evaluated with improvised ABCD attributes for the development of rule-based classifier as compared to the technique introduced by Mabrouk et al. [31]. For the estimation of ABCD attributes, significant improvisation has been introduced by incorporating the consideration of shape, brightness, and color asymmetry of the lesion along with spatial properties of dermoscopic structures. In this present study, to bring out considerable color information from the skin lesion area, a superpixel-based CIEM is introduced by considering the colors of the corresponding expert's annotated regions as reference. In comparison to the work [31], in the present study five differential structures have been identified as pigment network, dots, globules, structureless area and blue−white veil, for the evaluation of "D" score of ABCD rule of dermoscopy. The performance of the DermESy has been evaluated on 6669 dermoscopic images of various disease classes that justifies the robustness of the system.

The reported DermESy has been envisaged as a fully automated system irrespective of the datasets involved. It has been developed by considering the dermoscopic images of ISIC archive database and PH2 dataset. Therefore, a large number of samples with wide varieties of diseases have been incorporated for the identification of malignant and benign lesions. DermESy has automatically differentiated malignant, benign and suspicious categories of the sample dermoscopic images irrespective of the image source following the same steps and parameter settings. Acceptable performance indices for all the images from various resources with different image acquisition techniques affirmed the robustness of the system. Hence, it may be expected that the method can work with reasonable performance metrics irrespective of the datasets involved.

A graphical user interface of the developed expert system is shown in Fig. 5.15. Doctor can use this system to visualize the segmented lesion area, its border, color regions, and other differential structures for further assessment and monitoring of the disease.

5.5.6 Comparative analysis of dermoscopic findings of DermESy with expert's opinion

Employing the DermESy technique, ABCD attributes of the ABCD rule of dermoscopy have been estimated by finding different colors, structures, and quantification of morphological properties of the lesion. The presence of white (W), red (R), light brown (LB), dark brown (DB), blue−gray (BG), and black (B) colors helps to identify the malignant lesions at an early stage of the disease. Also, the structural asymmetry of the lesion is a very important parameter to identify the character of the lesion. Presence of different differential structures, predominantly the pigment network and the dots and globules identify the malignant lesions. In Fig. 5.16, a set of sample dermoscopic images have been shown along with their corresponding dermoscopic findings obtained using DermESy and expert dermatologist. The figure depicts the presence of all the six colors in the lesion area and illustrates the multiplicity of colors in malignant lesions compared to the benign lesions. The findings of color information have been verified with dermatologist's findings as shown in the figure. It justifies the correlation between the dermatologist opinion and findings of DermESy. It illustrates the presence of five differential structures obtained by DermESy. From the figure, it has been observed that structures extracted by the DermESy from most of the dermoscopic images have been confirmed by the expert. It also describes that the presence of

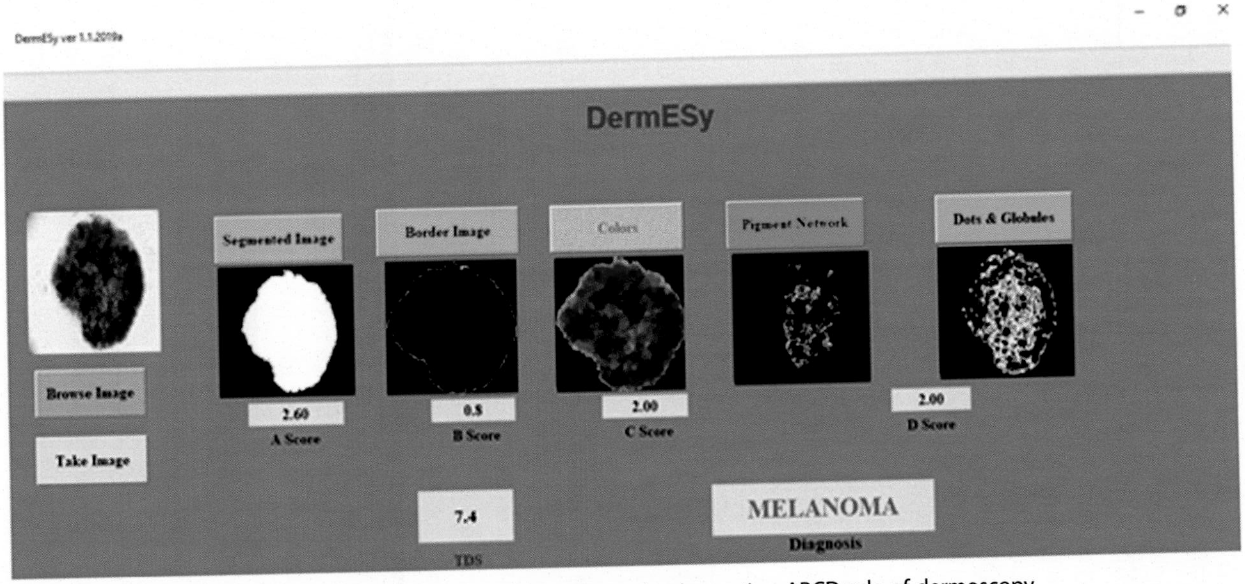

Figure 5.15 Graphical user interface of the DermESy, an expert system implementing ABCD rule of dermoscopy.

Sample Image	DermESy/ Expert	Asy	B	C						D				
				W	R	LB	DB	BG	B	PN	Dots	Globules	BWV	SL
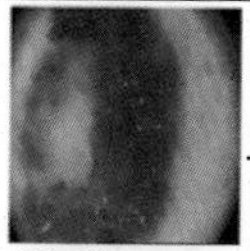 Malignant	DermESy	P	P	P	A	P	P	P	P	P	A	A	P	P
	Expert	P	P	P	A	P	P	P	P	P	P	A	P	P
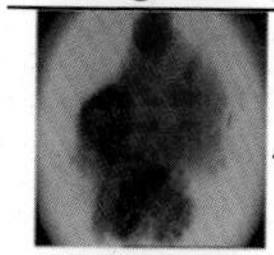 Malignant	DermESy	P	P	A	A	P	P	A	P	P	P	A	A	A
	Expert	P	P	A	A	P	P	A	A	P	P	A	A	A
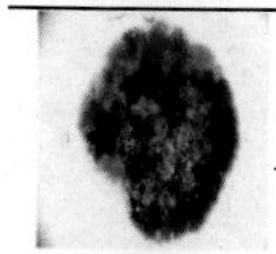 Malignant	DermESy	P	P	P	A	P	P	P	P	P	P	P	A	A
	Expert	P	P	P	A	P	P	P	P	P	P	P	P	P
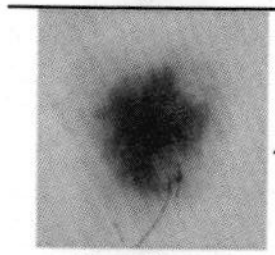 Benign	DermESy	A	A	A	A	P	P	A	A	P	P	A	A	A
	Expert	P	P	A	A	P	A	A	A	A	P	A	A	A
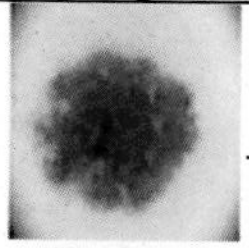 Benign	DermESy	A	A	A	A	P	P	A	P	P	A	A	A	A
	Expert	A	A	A	A	P	P	A	P	P	P	P	A	P

Figure 5.16 Identification of ABCD attributes from sample dermoscopic images using DermESy (*BWV*: Blue−white veil; *SL*: structureless area; *W*, white; *R*, red; *LB*, light brown; *DB*, dark brown; *BG*, blue−gray; *B*, black; *P*, present; *A*, absent; *PN*, pigment network).

pigment network and its spatial distribution has differentiated the malignant and benign lesions. Considering the spatial distribution of differential structures along the lesion, the ABCD rule has been improvised to evaluate TDS for the identification of the lesion. Morphological properties of skin lesions have been evaluated using DermESy as shown in Fig. 5.16. It depicts the lower asymmetry index and structural irregularity present in the lesion that leads to the identification of benign skin diseases. From the figure, it has been perceived that DermESy has identified some of the color and structural properties, which have not been detected from visual inspection by the dermatologist. Using DermESy, identification of various important dermoscopic findings and proper estimation of morphological properties of the lesion helps to determine the TDS more precisely for early-stage diagnosis of the diseases. Much better identification of ABCD attributes helps the dermatologist to correlate the dermoscopic findings and further analysis of the disease.

5.5.7 Performance comparison of DermESy with state-of-the-art automatic systems

In the present study, the dermoscopic images have been collected from ISIC challenge (2016, 2017, 2018) and PH2 dataset. Among 6669 dermoscopic images, 200 images are from PH2 dataset (160 benign and 40 malignant lesions) and remaining 6469 images are from ISIC challenge dataset. ISIC dataset comprises of dermoscopic images for 2016, 2017, and 2018 challenge. ISIC 2016 challenge dataset contains 1279 dermosopic images (900 training images and 379 testing images). In ISIC 2017 challenge dataset, additional 1696 dermoscopic images have been included. Remaining 3494 dermoscopic images collected for this study are from ISIC 2018 challenge dataset. For the characterization of the skin lesions using DermESy, ABCD attributes have been determined from each of the dermoscopic images to estimate the corresponding TDS. Depending on the TDS parameter value, the dermoscopic images have been classified as malignant, benign, and suspicious category. For TDS value less than 4.75, the lesions are classified as benign and for value more than 5.45, the lesions are classified as malignant. The TDS value within 4.75 and 5.45 corresponds to suspicious lesion category. Considering these parameter settings of the TDS score, the dermoscopic images of ISIC and PH2 datasets have been classified separately to compare the performance of DermESy with state-of-the-art techniques applied on the same datasets.

An integrated deep convolutional network-based approach has been introduced by Al-masni et al. [32] to classify skin lesions from ISIC challenge 2016 dataset. The reported deep convolutional network-based technique has obtained 81.79% classification accuracy for 379 test images. To compare the classification performance of the DermESy with the work by Al-masni et al. [32], the same set of dermoscopic images has been considered. Evaluating the TDS score from corresponding ABCD attributes from each of the 379 dermoscopic images, the skin lesions have been classified with

98.62% accuracy. The same work of Al-masni et al. [32] has tested the classification performance on ISIC 2017 dataset and obtained 81.57% accuracy. An ensemble deep convolutional neural networks has been used by Harangi [14] to classify the skin abnormalities from the available dermoscopic images of ISIC challenge 2017 dataset. The authors [14] have evaluated the performance on 600 testing images and obtained 86.60% classification accuracy. The work by Mahbod et al. [33] has proposed a combination of intra-architecture and interarchitecture network fusion using convolutional neural network for the identification of skin diseases on the same ISIC 2017 dataset. The authors have reported 87.70% classification accuracy by considering same set of images for testing purpose. The performance of the reported study has been compared with these state-of-the-art techniques, applied on same ISIC challenge 2017 dataset and has achieved 97.84% accuracy. A generative adversarial network based system has been employed by Qiu et al. [34] to classify the skin lesions from ISIC 2018 dataset. The work has achieved 95.20% classification accuracy by considering 2003 number of test images of ISIC 2018 dataset. The work by Al-masni et al. [32] has also evaluated the classification performance on the same 2003 dermoscopic images of ISIC challenge 2018 dataset and has reported 89.28% accuracy. The reported study has also been tested on the same 2003 number of dermoscopic images and has obtained the classification accuracy of 97.02%. Table 5.6 reveals that the reported expert system has outperformed the state-of-the-art methodologies for the identification of skin diseases with higher degree of accuracy on ISIC challenge 2016, 2017, and 2018 datasets.

In Table 5.7, the performance of the reported DermESy has been compared with some of the recently published works for skin disease classification on publicly available PH2 dataset. A diagnostic system for melanoma identification using ABCD rule of dermoscopy has been reported by Zaqout [35]. The author [35] has reported 90% accuracy

Table 5.6 Classification performance comparison of the proposed system with state-of-the-art techniques on ISIC challenge dataset.

Dataset	Method	Classification accuracy (%)
ISBI/ISIC 2016 dataset (900 training images, 379 testing images)	Al-masni et al. [32]	81.79
	Present study	98.62
ISBI/ISIC 2017 dataset (2000 training images, 600 testing images)	Al-masni et al. [32]	81.57
	Harangi [14]	86.60
	Mahabod et al. [33]	87.70
	Present study	97.84
ISBI/ISIC 2018 dataset (total 10,015 number of images, 2003 images for testing)	Qin et al. [34]	95.20
	Al-masni et al. [32]	89.28
	Present study	97.02

ISBI, International symposium on biomedical imaging; ISIC, International Skin Imaging Collaboration.

Table 5.7 Classification performance comparison of the proposed system with state-of-the-art techniques on PH2 dataset.

Dataset	Method	Classification accuracy (%)
PH2 dataset (200 images for testing)	Zaqout [35]	90.00
	Ghalejoogh et al. [36]	96.00
	Hu et al. [37]	91.90
	Gulati and Bhogal [38]	94.50
	Present study	98.48

for melanoma identification from 200 dermoscopic images of PH2 dataset, which has not been verified with an expert's opinion, in contrast to this present study. A stacking ensemble method based on the meta-learning algorithm has been proposed by Ghalejoogh et al. [36] for the classification of skin abnormalities. The reported work [36] has classified the skin lesions from the dermoscopic images of PH2 dataset with 96% classification accuracy. The research work by H et al. [37] has introduced bag-of-feature model using feature similarity measurement technique based on codebook learning algorithm for the classification of melanoma from the dermoscopic images of PH2 dataset. In this work [37], 91.90% accuracy has been attained for the classification of skin lesions. A support vector machine classifier based skin disease identification technique has been proposed by Gulati and Bhogal [38] and obtained 94.50% classification accuracy. To compare the classification performance of the reported expert system with state-of-the-art techniques on PH2 dataset, TDS value has been estimated from each of the 200 dermoscopic images. Based on the TDS value, the skin lesions have been identified with 98.48% accuracy. Table 5.7 portrays that the reported ABCD rule-based expert system has achieved an acceptable classification accuracy for skin disease diagnosis from PH2 dataset. The results have described a significant improvement in classification performance compared to the recently published works on same dataset.

5.5.8 Comparative analysis of the diagnosis of skin lesions using DermESy and expert dermatologist

The performance of the expert system is compared with the expert's diagnosis for establishing the robustness of the algorithms. Dermatologist of this research group has identified the malignant and benign lesions from the dermoscopic image set. Dermatologist has determined the corresponding TDS of each lesion based on their clinical criteria described by ABCD attributes. Fig. 5.17 has depicted the diagnosis of skin lesions using DermESy and the dermatologist involved in this study. The corresponding TDS score evaluated by the dermatologist and DermESy have been given for the ready reference. From the figure, it has been observed that improvisation of ABCD attributes in DermESy helps to categorize suspicious lesion into malignant or

Sample Images	Diagnosis by DermESy		Diagnosis by Dermatologist	
	TDS	Category	TDS	Category
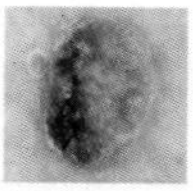	6.50	Malignant	6.00	Malignant
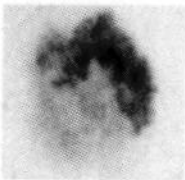	5.85	Malignant	5.05	Suspicious
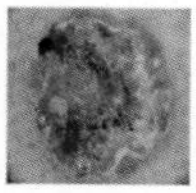	5.60	Malignant	4.90	Suspicious
	4.25	Benign	5.00	Suspicious
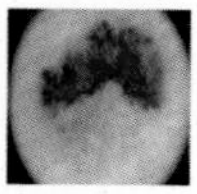	6.25	Malignant	5.85	Malignant
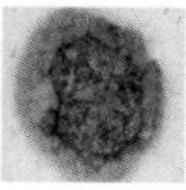	4.00	Benign	4.25	Benign

Figure 5.17 Comparative analysis of the diagnosis of skin lesions using DermESy and expert dermatologist (*TDS*, Total dermoscopic score.)

benign class. Similarly, identification of a set of lesions using DermESy is in line with the diagnosis of the dermatologist. Therefore, the diagnosis of the skin lesion by DermESy can be used as a second opinion apart from the expert's opinion for further decision-making.

5.6 Conclusion

The study introduces an expert system for skin disease diagnosis. The DermESy has implemented the standard ABCD rule of dermoscopy with significant improvisations. For the estimation of lesion asymmetry score, shape, color, and brightness asymmetry have been evaluated. To incorporate dermatologist's expertise, the color information extraction algorithm has been introduced by considering the expert's annotated color information as reference. Dermatologists consider the spatial information of the differential structures to differentiate the malignant and benign lesions. This expertise has been introduced to improvise the ABCD criteria by considering spatial information of the structures to determine the "D" score. The rule base of the system has been developed by incorporating the expert dermatologists' knowledge for improved early-stage diagnosis of the disease and further decision-making. The incorporation and implementation of ABCD rule of the skin disease diagnosis has provided an in-depth visualization and quantification of dermoscopic findings for further course of action. The performance of the DermESy has been compared with the expert dermatologist's diagnosis on a large set of dermoscopic images. DermESy has identified the malignant and benign lesions with 97.86% sensitivity in comparison with the dermatologist's diagnostic accuracy of 90.91%. Since the performance of the proposed system is evaluated on a large set of images and the diagnosis is also compared with that by a dermatologist, the robustness of the system is ensured. Therefore, DermESy could be an effective tool not only for an expert dermatologist but also for the radiologist or a general physician to evaluate competently the state of the lesion area for further treatment. In future, this DermESy system can be improved for the development of more sophisticated software to provide second opinion to the dermatologists for improved and uniform diagnosis of the disease at an early stage. This study can be extended with required modifications and inclusion of diameter estimation of the lesion for the differentiation of malignant and benign lesions from clinical or macroscopic images. Advanced digital signal processing tools can be introduced to implement this study for remote monitoring of the skin diseases. A microcontroller based integrated system can also be developed implementing the reported DermESy system for the construction of a hand-held device for skin disease monitoring.

References

[1] R. Marks, An overview of skin cancers: incidence and causation, Cancer Suppl. 75 (S2) (1995) 607–612.

[2] H. Pehamberger, M. Binder, A. Steiner, K. Wolff, In vivo epiluminescence microscopy: improvement of early diagnosis of melanoma, J. Invest. Dermatol. 100 (1993) 356S–362S.

[3] F. Xie, H. Fan, Y. Li, Z. Jiang, R. Meng, A. Bovik, Melanoma classification on dermoscopy images using a neural network ensemble model, IEEE Trans. Med. Imaging 36 (3) (2017) 849–858.

[4] H.P. Soyer, G. Argenziano, S. Chimenti, V. Ruocco, Dermoscopy of pigmented skin lesions, Eur. J. Dermatol. 11 (2001) 270–276.

[5] J. Kawahara, S. Daneshvar, G. Argenziano, G. Hamarneh, 7-Point checklist and skin lesion classification using multi-task multi-modal neural nets, IEEE J. Biomed. Health Inform. (2018). Available from: https://doi.org/10.1109/JBHI.2018.2824327.

[6] R. Kasmi, K. Mokrani, Classification of malignant melanoma and benign skin lesions: implementation of automatic ABCD rule, IET Image Process. 10 (6) (2016) 448–455.

[7] I. Maglogiannis, K. Delibasis, Enhancing classification accuracy utilizing globules and dots features in digital dermoscopy, Comput. Method Prog. Biomed. 118 (2015) 124–133.

[8] A. Sáez, J. Sánchez-Monedero, P.A. Gutiérrez, C. Hervás-Martínez, Machine learning methods for binary and multiclass classification of melanoma thickness from dermoscopic images, IEEE Trans. Med. Imaging 35 (4) (2016) 1036–1045.

[9] R.B. Oliveira, A.S. Pereira, J.M.R.S. Tavares, Skin lesion computational diagnosis of dermoscopic images: ensemble models based on input feature manipulation, Comput. Method Prog. Biomed. 149 (2017) 43–53.

[10] F. Nachbar, et al., The ABCD rule of dermatoscopy: high prospective valie in the diagnosis of doubtful melanocytic skin lesions, J. Am. Acad. Dermatol. 30 (4) (1994) 551–559.

[11] M.A. Wahba, A.S. Ashour, Y. Guo, S.A. Napoleon, M.M.A. Elnaby, A novel cumulative level difference mean based GLDM and modified ABCD features ranked using eigenvector centrality approach for four skin lesion types classification, Comput. Method Prog. Biomed. 165 (2018) 163–174.

[12] Y. Yuan, Y. Lo, Improving dermoscopic image segmentation with enhanced convolutional–deconvolutional networks, IEEE J. Biomed. Health Inform. 23 (2) (2019) 519–526.

[13] I. Guyon, S. Gunn, M. Nikravesh, L.A. Zadeh, Feature Extraction Foundations and Applications, vol. 207, Springer, 2006.

[14] B. Harangi, Skin lesion classification with ensembles of deep convolutional neural networks, J. Biomed. Inform. 86 (2018) 25–32.

[15] I. González-Díaz, DermaKNet: incorporating the knowledge of dermatologists to Convolutional Neural Networks for skin lesion diagnosis, IEEE J. Biomed. Health Inform. 23 (2) (2019) 547–559 (2018).

[16] P. Soille, Morphological Image Analysis Principles and Applications, second ed., Springer, 2004.

[17] R. Achanta, A. Shaji, K. Smith, A. Lucchi, P. Fua, S. Süsstrunk, SLIC superpixels compared to state-of-the-art superpixel methods, IEEE Trans. Patt. Anal. Mech. Intell. 34 (11) (2012) 2274–2281.

[18] C. Barata, J.S. Marques, J. Rozeira, A system for the detection of pigment network in dermoscopy images using directional filters, IEEE Trans. Biomed. Eng. 59 (10) (2012) 2744–2754.

[19] S. Pathan, K.G. Prabhu, P.C. Siddalingaswamy, A methodological approach to classify typical and atypical pigment network patterns for melanoma diagnosis, Biomed. Signal Proc. Control. 44 (2018) 25–37.

[20] M. Katz, Fractals and the analysis of waveforms, Comput. Biol. Med. 18 (3) (1988) 145–156.

[21] D. Gutman, et al., Skin lesion analysis toward melanoma detection: a challenge at the international symposium on biomedical imaging (ISBI) 2016, hosted by the international skin imaging collaboration (ISIC), 2016 [Online] Available: <https://arxiv.org/abs/1605.01397>.

[22] T. Mendonça, P.M. Ferreira, J. Marques, A.R.S. Marcal, J. Rozeira, PH2—a dermoscopic image database for research and benchmarking, in: 35th International Conference of the IEEE Engineering in Medicine and Biology Society, July 3–7, 2013, Osaka, Japan.

[23] M.A. Al-masni, M.A. Al-antari, M.-T. Choi, S.-M. Han, T.-S. Kim, Skin lesion segmentation in dermoscopy images via deep full resolution convolutional networks, Comput. Method Prog. Biomed. 162 (2018) 221–231.

[24] L. Bi, J. Kim, E. Ahn, A. Kumar, D. Feng, M. Fulham, Step-wise integration of deep class-specific learning for dermoscopic image segmentation, Pattern Recognit. 85 (2018) 78–89.

[25] P. Shan, Y. Wang, C. Fu, W. Song, J. Chen, Automatic skin lesion segmentation based on FC-DPN, Comput. Biol. Med. 123 (2020) 103762.

[26] F. Xie, J. Yang, J. Liu, Z. Jiang, Y. Zheng, Y. Wang, Skin lesion segmentation using high-resolution convolutional neural network, Comput. Method Prog. Biomed. 186 (2020) 105241.

[27] D. Piccolo, G. Crisman, S. Schoinas, D. Altamura, K. Peris, Computer-automated ABCD vs dermatologists with different degrees of experience in dermoscopy, Eur. J. Dermatol. 24 (2014) 477–481.

[28] N. Smaoui, S. Bessassi, A developed system for melanoma diagnosis, Int. J. Comput. Vis. Signal Process. 3 (2013) 10–17.

[29] S.J. Birkenfeld, M.J. Tucker-Schwartz, R.L. Soenksen, A.J. Avilés-Izquierdo, B. Marti-Fuster, Computer-aided classification of suspicious pigmented lesions using wide-field images, Comput. Methods Prog. Biomed. 195 (2020) 105631.

[30] G.A. Isasi, G.B. Zapirain, M.A. Zorrilla, Melanomas non-invasive diagnosis application based on the ABCD rule and pattern recognition image processing algorithms, Comput. Biol. Med. 41 (2011) 742–755.

[31] S.M. Mabrouk, Y.A. Sayed, M.H. Afifi, A.M. Sheha, A. Sharwy, Fully automated approach for early detection of pigmented skin lesion diagnosis using ABCD, J. Healthc. Inform. Res. (2020). Available from: https://doi.org/10.1007/s41666-020-00067-3.

[32] M.A. Al-masni, D. Kim, T. Kim, Multiple skin lesions diagnostics via integrated deep convolutional networks for segmentation and classification, Comput. Method Prog. Biomed. 120 (2020) 105351.

[33] A. Mahbod, G. Schaefer, I. Ellinger, R. Ecker, A. Pitiot, C. Wang, Fusing fine-tuned deep features for skin lesion classification, Comput. Med. Imaging Graph. 71 (2019) 19–29.
[34] Z. Qin, Z. Liu, P. Zhu, Y. Xue, A GAN-based image synthesis method for skin lesion classification, Comput. Method Prog. Biomed. 195 (2020) 105568.
[35] I. Zaqout, "Diagnosis of Skin Lesions Based on Dermoscopic Images Using Image Processing Techniques," IntechOpen, http://doi.org/10.5772/intechopen.88065, 2019.
[36] G.S. Ghalejoogh, H.M. Kordy, F. Ebrahimi, A hierarchical structure based on Stacking approach for skin lesion classification, Expert Syst. Appl. 145 (2020) 113127.
[37] K. Hu, X. Niu, S. Liu, Y. Zhang, C. Cao, F. Xiao, et al., Classification of melanoma based on feature similarity measurement for codebook learning in the bag-of-features model", Biomed. Signal Process. Control. 51 (2019) 200–209.
[38] S. Gulati, R.K. Bhogal, Classification of melanoma from dermoscopic images using machine learning, Smart Innov. Syst. Technol. 159 (2020) 345–354.

CHAPTER 6

Conclusions and future scope of work

6.1 Conclusions

The book addresses the challenges for the development of a computer aided skin disease identification system for early, accurate, and uniform evaluation of the abnormalities for further treatment, monitoring, and prevention of the disease. As the dermatologists examine the dermoscopic images by visual inspection, the skin disease identification system has been developed here using dermoscopic images of various skin diseases. Four different aspects of skin disease monitoring systems have been addressed in this book. These aspects are: (1) removal of noise and artifacts from the dermoscopic images employing preprocessing techniques, (2) segmentation of skin lesion to segregate the affected area and monitor further spreading of the disease, (3) development of feature extraction tools using advanced signal processing techniques to quantify significant representative features from the skin lesion images, and (4) development of classification models to differentiate closely similar skin diseases from the selected demarcating features using efficient feature selection and classification technique. Dermatologists follow standard ABCD rule of dermoscopy for the differentiation of malignant and benign lesions. In this book, an expert system replicating the ABCD rule of dermoscopy is developed to get a second opinion for the dermatologist and general physician, for early and accurate diagnosis of the disease. In the preprocessing stage, the noise due to improper lighting condition and hair artifacts are removed from the dermoscopic images. The uneven illumination of light during image acquisition and other parameters of the dermoscopes introduce uneven intensity distribution across the lesion area. The dermoscopic images of the skin lesions in hairy regions of the human body introduce hair artifacts that may lead to the improper visualization of the affected area and identification of significant information. Due to the presence of these noises and artifacts, segregation of actual affected region from the image becomes very much challenging. In this book, mathematical morphology is used to develop morphological filters for the detection and extraction of hair artifacts. The primary element of morphological operation is the structuring element. Selection of a proper structuring element for a specific purpose is a very critical criterion. In a study in this book, a circular kernel has been introduced to develop morphological filters. Morphological bottom hat filter with varying size of the structuring element is introduced to extract the hair like objects having variable width and darker intensity than their surroundings. This technique has also eliminated the problem of existing hair

Recent Trends in Computer-aided Diagnostic Systems for Skin Diseases
DOI: https://doi.org/10.1016/B978-0-323-91211-2.00006-8

shadows on lesion area. Elimination of noise and artifacts from the original images helps in appropriate segmentation of lesion area for further monitoring of the disease.

Segmentation of the affected area from the dermoscopic images is a challenging task due to the complex and irregular structure of the lesion. Proper segmentation of the lesion area helps to determine the morphological properties of the lesion and to monitor further spreading of the disease. In this book, a wide varieties of skin diseases having different structural complexities has been considered. Malignant melanoma has a very complex and irregular structure compared to the benign nevus. Segmentation of basal cell carcinoma (BCC) lesion is very much challenging due to its shape and scattered properties. Elimination of the hair artifacts from the affected area dispenses with the unnecessary oversegmentation of the object, that may lead to the improper diagnosis of the disease. To segregate the region of interest from the dermoscopic images, mathematical morphology has been widely explored in this book. As it deals with the morphological properties of the object, mathematical morphology has been chosen as the tool for the development of skin disease segmentation algorithm. Considering the structural properties of the skin lesions, circular kernel is employed as the structuring element for the morphological operations. Morphological gradient operation is introduced to determine the object of interest depending on its structural properties. After the segmentation of the lesion area, morphological gradient-based single-pixel border-detection technique is introduced for further determination of the border irregularity of the lesion. A simple and efficient technique having low computation and time complexity is introduced to segment various skin lesions with different structural complexities. To evaluate the performance of the segmentation algorithm, the pixel level sensitivity, the specificity, the accuracy, along with the image similarity measuring indices as Jaccard similarity index (JSI) and Dice similarity coefficient (DSC) are determined with respect to the ground truth images. Mathematical morphology-based segmentation methodology has segmented the lesion area with 91.72% sensitivity, 97.88% specificity, and 95.21% accuracy. The reported study has achieved acceptable similarity indices of 85.62% JSI and 91.42% DSC, thus exhibiting acceptable performance indices that outperform other state-of-the-art techniques.

For the development of computer aided skin disease monitoring system, extraction of significant features is very much important. Qualitative analysis of the morphological, texture and color properties of the skin lesion make it difficult for the dermatologists in early-stage detection and prevention of the disease. In this book, digital signal processing tools have been developed for the quantification of dermatological properties of skin lesions. To describe the morphological properties of the lesion, a set of statistical features such as area, perimeter, circularity have been determined from the segmented lesion area. Structural irregularity of the lesion is a demarcating feature for the differentiation of skin diseases. Single-valued Haussdorff fractal dimension is estimated to quantify the structural complexity from the border detected images. For more precise evaluation, a wavelet-based border irregularity measurement technique is introduced. Employing wavelet transform, border series has been decomposed into low and high frequency regions to determine certain changes in the lesion structure. Therefore determination of Higuchi and Katz's fractal dimension from each of the decomposed signal provides more detailed description of border irregularity. Texture is most subjective in nature and very difficult to identify and distinguish by visual inspection. Thus quantification of textural complexity of the lesion is very much challenging for wide varieties of skin lesions. Some efficient texture analysis techniques have been developed using advanced signal processing tools along with popular cooccurrence matrix-based techniques. A fractal-based regional texture analysis (FRTA) technique is introduced to identify the localized textural complexity of the skin lesion. The algorithm decomposes the lesion area into significant smaller textural regions, having significant intensity variations and fractal dimension has been estimated from each of the region. Depending on the fractality, the larger regions are further decomposed into smaller subregions. However, the FRTA algorithm only considers the intensity variation along the lesion area, without considering the spectral information. To incorporate spectral information, wavelet packet fractal texture analysis (WPFTA) is introduced. Wavelet transform decomposes the dermoscopic image into approximate and detail coefficients in spatial—spectral domain. Employing wavelet packet decomposition technique, subsequent decomposition of approximate and detail coefficients provides more finer details of the textural pattern. However, the single-valued fractal dimension does not describe the entire textural irregularity. To quantify the entire textural information, fractal descriptor is constructed for each of the decomposed images. Therefore to analyze detailed texture feature, WPFTA has been deployed. In visual inspection, the dermatologist explores different color or dermatological pattern to identify various skin abnormalities. Presence of such information bearing regions ensures proper diagnosis of the disease. To incorporate the dermatologist's findings to develop the skin disease diagnosis system, cross-correlation-based technique is introduced. Utilizing the feature descriptors (regions) annotated by the expert dermatologist, the cross-correlation technique has determined the significant regions from the dermoscopic images based on the degree of commonality between the feature descriptors and input images. Spatial and spectral features extracted from the cross-correlogram describe the regional color or texture features. An unsupervised learning-based feature extraction technique is deployed by using sparse auto-encoder, to extract reduced set of distinguishable features from the dermoscopic images. Using stacked auto-encoder model, the learned weight values of the neural network are considered as reduced feature set. For the color feature extraction, superpixel-based regional color information extracting technique is introduced. In this book, different feature extracting tools have been discussed for the extraction of morphological, texture, and color features from the dermoscopic images.

After the feature extraction stage, the skin diseases are categorized into appropriate disease classes by employing a classification technique. For improved classification, the selection of demarcating features has significant contribution. In this book, support vector machine-based recursive feature elimination technique is used to obtain the reduced set of important features. When the number of features is significantly increased, the possibility of the presence of correlated features are increased. To address this issue, an automatic correlation bias reduction technique is deployed to obtain demarcating features from the large feature set. Incorporating different feature selection algorithms, binary, and multiclass classification techniques are introduced. Employing support vector machine (SVM)-based binary classification techniques, melanoma and benign lesions are differentiated. To classify these two disease classes, a reduced set of features is obtained by employing correlation bias reduction-based support vector machine-based recursive feature elimination (SVM-RFE) technique. Correlated features have been eliminated in an automatic manner depending on the classification performance. This reported study has achieved 97.63% sensitivity, 100% specificity and 98.28% correct classification accuracy. In this book, SVM-based binary classification technique has outperformed the performance of the existing methods reported in literature. A multiclass ensemble classification model is introduced to differentiate three skin disease classes as melanoma, nevus, and BCC, employing SVM-RFE feature selection technique. This stage wise classification technique has identified these three disease classes with 98.48%, 99.39%, 99.63% sensitivity, 98.28%, 91.07%, 100% specificity and 98.99%, 97.54%, 99.65% accuracy for melanoma, nevus, and BCC images respectively. This multiclass classification approach helps to identify the misclassified images at each stage of the classifier model for further analysis. Multilabel ensemble multiclass classification technique is introduced to differentiate four skin disease classes as melanoma, nevus, BCC, and seborrheic keratosis (SK) images. The four class classification model incorporating feature selection technique is deployed to categorize the diseases into melanocytic and nonmelanocytic category and subclasses. The reported study has achieved sensitivities of 98.76%, 99.01%, 98.87%, and 99.41% for melanoma, nevus, BCC, and SK diseases, respectively.

The final chapter of this book addresses the development of an expert system, implementing ABCD rule of dermoscopy to differentiate benign and malignant skin lesions. This expert system, namely DermESy is developed to assist the expert dermatologist in further decision making towards the early-stage diagnosis of the disease. Here, the gold standard rule of dermoscopy (ABCD rule) is replicated with significant improvisation using if—then rule, addressing all the four aspects. In DermESy, an explanatory subsystem is also introduced with decision making block to correlate the diagnosis with expert's opinion based on proper visualization. According to total dermoscopic score, DermESy has differentiated benign and malignant lesions with 97.69% sensitivity, 97.97% specificity, and 97.86% accuracy.

6.2 Future scopes

The results presented in this book can provide a basis for further research in various directions. Considering the advancement in signal processing techniques, some of the future scopes of research have been mentioned here:

- Studies in this book addresses the identification of skin diseases, one of most abundantly found ailments affecting human population of the world. To carry out these studies in this book, dermoscopic images have been collected from different freely available databases of various universities or hospitals of the skin cancer prone countries of the world. However, in Indian scenario, skin cancer is very rare compared to the other skin abnormalities. Therefore the ideas of this book can be used for the differentiation of some common skin diseases from their close proximities like Psoriasis, Tinea, etc.
- In these present studies, microscopic images of skin lesions have been considered to identify the skin diseases. Identification of these diseases from macroscopic or infrared images can explore a new segment of research. Macroscopic image provides significant color and texture information of the skin lesion and can be acquired using standard digital cameras. Some of the detailed information can be obtained from the lesion images beyond the visual spectrum, acquired using infrared cameras. Hyperspectral camera can also be used to acquire the images in wide spectral range, to yield wide varieties of information. The advanced signal processing tools can be developed for the extraction of features from these images of various spectral regions.
- An indigenous, microcontroller-based integrated system can be developed in conjunction with an image acquisition module and subsequent feature extraction and classification module. Practical implementation of this system may lead to the development of handheld device for condition monitoring of skin abnormalities.
- Condition monitoring of skin diseases from a remote location is an emerging topic of research. For the development of such remote monitoring system, the acquired images can be transmitted to a certain location for identification using a dermatologist's knowledge-based automatic system. This will help to assess a wide range of patient in reduced time. This computer aided diagnosis can be considered as a second opinion for the general physicians, without having expertise in that domain for accurate and early detection of the disease. Therefore one of the most important scopes of further research is remote monitoring of wide varieties of skin diseases employing advance signal processing tools.

The field of study of this present book is chosen as the abnormality detection of skin lesions. However, the methods and tools which are developed in this book are generic in nature. Hence the developed tools can be applied for analogous problems in the domain of biomedical systems as a part of future work.

Index

Note: Page numbers followed by "*f*" and "*t*" refer to figures and tables, respectively.

D

H

I

J

K

L

M

Printed in the United States
by Baker & Taylor Publisher Services